TREASURES OF THE EARTH

TREASURES OF THE EARTH

Need, Greed, and a Sustainable Future

Saleem H. Ali

Yale University Press
New Haven and London

"To Crystallization," from *Collected Poems, 1953–1993*, by John Updike, copyright © 1993 by John Updike. Used by permission of Alfred A. Knopf, a division of Random House, Inc.

Copyright © 2009 by Saleem H. Ali.
All rights reserved.
This book may not be reproduced, in whole or in part, including illustrations, in any form (beyond that copying permitted by Sections 107 and 108 of the U.S. Copyright Law and except by reviewers for the public press), without written permission from the publishers.

Set in Galliard type by Westchester Book Group.

Printed in the United States of America.

Library of Congress Cataloging-in-Publication Data
Ali, Saleem H., 1973–
  Treasures of the earth : need, greed, and a sustainable future / Saleem H. Ali.
    p. cm.
  Includes bibliographical references and index.
  ISBN 978-0-300-14161-0 (alk. paper)
    1. Consumption (Economics)—Environmental aspects. 2. Consumption (Economics)—Moral and ethical aspects. 3. Sustainable development. 4. Environmental policy. 5. Natural resources—Environmental aspects. 6. Raw materials—Environmental aspects. 7. Conservation of natural resources. I. Title.
  HC79.C6A42 2009
  333.7—dc22

                                                      2009018208

A catalogue record for this book is available from the British Library.

This paper meets the requirements of ANSI/NISO Z39.48-1992 (Permanence of Paper).

10 9 8 7 6 5 4 3 2

For Shahmir and Shahroze
Needful treasures of my life

Look to the rock from which you were hewn
And to the quarry from which you were dug
—Isaiah 51:1

CONTENTS

Preface ix

Introduction: Alchemy of a Material World 1

PART I: THE PLEASURE OF TREASURE

1. What Lies Beneath
   The Material and Mystical Origins of Mineral Wealth 17

2. Creating Value
   The Endurance of Precious Jewels 39

3. The Rush Factor
   Tracing the Mineral Roots of Global Power Through Gold, Coal, and Oil 62

PART II: TOIL AND TREASURE

4. The Darker Side of Fortune
   The Psychology of Treasure Dependence 89

5. Curing the Resource Curse
   Minerals and Global Development 110

6. The Spoils of the Earth
   The Ecological Toll of Extractive Excitement 132

PART III: MEASURE FOR MEASURE

7. Destination Cradle
   The Quest for Material Cycling   161

8. The Restoration Ethic
   Hurting and Healing Ecosystems   185

9. Embracing the Treasure Impulse
   Toward Cautionary Creativity   207

   Epilogue: Embracing Uncertainty   235

   Appendix
   Notable Minerals and Their Human Uses   239

   Notes   245

   Index   278

PREFACE

Would the world be a better place if human societies were somehow able to curb their desires for material goods? Such a simple question prompted the writing of this book. For much of my career, I have tried to reconcile technical indicators of environmental impact with societal values. As an environmental planner by training, I have naturally been focused on the future. As a conflict analyst, however, I have always felt anchored to the past in trying to understand the convergence of factors that lead to community tensions. Often, larger societal conflicts in troubled times boil down to a series of internal conflicts among individuals about what degrees of consumption are essential and which are superfluous in the context of environmental challenges. Yet much scholarship tends to be very polarized and linear in its causal explanations about the linkages between human consumption, well-being, and the environment. For the public, the result is a continuing standoff that leads to policy paralysis and inaction. At its core, this is a tension between human "needs" versus "wants." For environmentalists, self-denial of human wants is a virtue, and a minimalist lifestyle is the road to salvation. For market enthusiasts, willingness to pay is a sufficient consideration for any noncriminal resource or product, regardless of whether it is "needed" or not.

Although efficiency of resource usage must always be a primary goal, this book attempts to present a humanitarian case for resource extraction that goes beyond meeting human needs. At the heart of the matter is our ability to understand the world's dependence on nonrenewable resources—namely minerals on which all human societies have always depended as primary raw materials. I define minerals for the purposes of this book as anything that is mined from the earth, which reflects the etymological root of the word. This dependence is usually presented only in terms of consumption, but the more salient dependence lies in areas of production. The distribution of the world's mineral resources is an accident of geography, but it has considerable causal power in determining the trajectory of human societies.

Understanding the chemical and physical constraints of the elements was also an important consideration for me in preparing this narrative. Harmonizing many diffuse areas of knowledge about our relationship to materials and the impact of those relationships on the environment requires an unusual breadth of inquiry. Although my academic credentials are fairly eclectic, traversing fields as broad as economic history, psychology, and chemistry has required a lot of self-learning. With current social networking technologies, I hope to continue the conversation of this book with readers around the world and develop new ideas for best harnessing our treasure impulse.

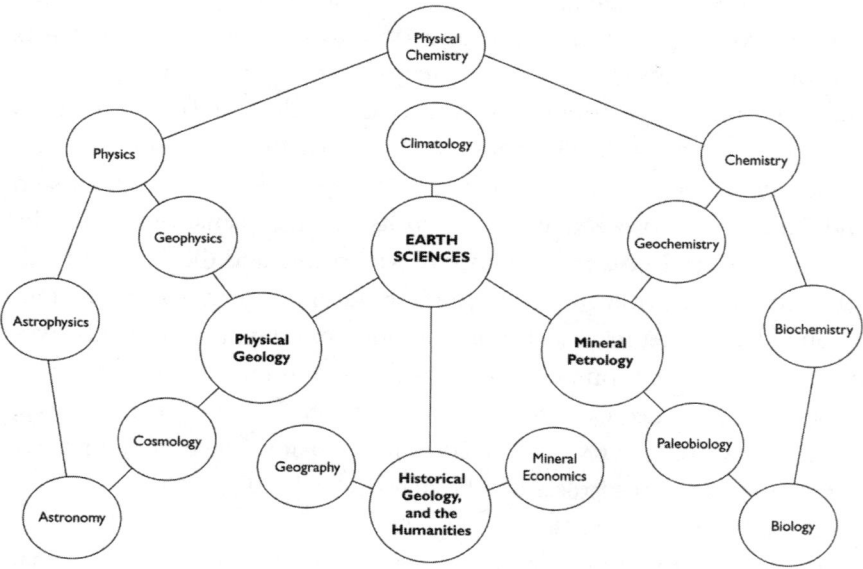

FIGURE A. Harmonizing the earth sciences across disciplines (Adapted from a diagram at the National Museum of Scotland, Edinburgh)

I was alerted to the confluence of various fields of natural inquiry around minerals while visiting the National Museum of Scotland in Edinburgh many years ago, where one finds an intriguing diagram outside the hall of minerals. Drawing on my roots as a chemistry major, much of the material in this book is based on an appreciation for the elements of the earth. In teaching large undergraduate courses on environmental studies, I have been alarmed by how few of my students are inclined to understand fundamental chemical principles. Often, students are also not willing to be critical thinkers about environmental challenges and would rather be entrenched in normative positions about issues like consumerism or globalization. The general public shows a similar reluctance to wrestle intellectually with tough issues. This book is an attempt to stimulate critical thinking about the environment without disparaging well-intentioned efforts by activists on either side of the political spectrum.

Although environmentalism had strong Malthusian roots for much of the twentieth century, the appeal of population as the central variable in activism began to fade after the Earth Summit in 1992, and the emphasis shifted instead to consumption. It became more acceptable to self-flagellate affluent consumer lifestyles in our own backyard than to cast aspersions concerning demographic growth in developing countries. In more technical terms, the famed "IPAT equation,"

$$\text{Environmental Impact (I)} = \text{Population} \times \text{Affluence} \times \text{Technology},$$

which graces most environmental textbooks, needed to have its weighting of variables reconsidered. Environmental activism has of late shifted its focus more to the affluence variable as a result of political pressures. Yet the fundamental challenge of resource utilization still has the population variable playing a pivotal role at multiple levels. People are generally living longer, worldwide, and scientists are still trying to search for increased longevity through a biochemical reversal of the aging process. Even if we could persuade people to have fewer children, we are faced with the challenge of an aging population and the ethical imperative to respect life.

Furthermore, the technology variable is often treated like a residual unknown in standard economic models and is scarcely part of the process of environmental planning. Even when technology is presented as an active variable, the difficulty in measuring the diffuse applications of scientific advancement often leads to a polarization between Cornucopian and Cassandran views of human development potential. This book attempts to grapple with the rather imponderable aspect of technologically driven modernity, which requires us to stimulate human creativity and innovation in ways that have previously been exercised only in the context of earthly exploration.

Another major motivation for this book has been the clash of conversations on resource dependency due to global geopolitics. Since 9/11, there has been a proliferation of books that focus on oil imports and the consequences for oil dependence and "petropolitics." Similarly, there has been much interest in the linkage between diamond mining and conflict, popularized by such movies as *Blood Diamond*. However, these writings and productions do not approach the topic from an integrated environmental and social perspective—choosing instead to provide a linear explanation for poor governance by simplistically focusing on particular minerals. Such narratives present a stylized view of resource extraction, lacking a broader appreciation for the relatively few livelihoods that exist in many of these areas, which must be considered when evaluating particular consumption decisions. The analysis is also historically selective and neglects larger questions of income inequality between rich and poor countries. Clearly, development assistance does not have a great record of alleviating poverty on its own. Yet the more consequential question is, if minerals were not extracted to jump-start economies in many countries, would alternative livelihoods be viable?

This book has an ambitious goal to reconfigure the conversation on environmental behavior through a recognition that consumption is not necessarily a "sin" (as many environmentalists may suggest), or a "virtue" (as most economic models assume). Rather, the impulse that leads to consumption behavior has important repercussions for livelihoods all over the world, and these must be considered in effective environmental planning. As for the initial question that began this preface, readers will find that the answer is more nuanced than expected.

Finishing a book of this scope is a team effort at many levels. My colleagues and students at the University of Vermont, Brown University, Griffith University in Australia, and the University for Peace in Costa Rica have provided intellectual nourishment for this project during the research and writing phase. In particular, my graduate students Mary Ackley, Samir Doshi, Sally Dickinson-Deleon, Ganlin Huang, and Anthony McInnis helped with research at many phases. My undergraduate research assistant Kathryn Romelczyk persistently helped with securing permissions for epigrams and quotations.

Scholars from across the world with whom I have engaged in constructive arguments and who have shared important data with me include Tom Graedel, Sharman Haley, Anthony Hodge, Kuntala Lahiri-Dutt, Estelle Levin, Jonathan Isham, Philippe LeBillon, Reid Lifset, Satoshi Murao, Ciaran O'Faircheallaigh, John Todd, and Natalia Yakovleva. The publishing team, including my agent Gillian MacKenzie, the acquisitions editor at Yale University Press Jean Thompson-Black, and editorial assistants Matthew Laird and Joseph Calamia were encouraging and meticulous.

My wife, Maria; our children, Shahmir and Shahroze; my mother, Parveen; and my sisters Irfana and Farzana have shown tremendous patience in tolerating long work hours and prolonged absences for fieldwork. Continuing gratitude is owed to my father-in-law, Saleem Kashmiri, who also taught me English literature in a Pakistani high school almost two decades ago, and instilled in me a love of reading and writing with care and constancy. Throughout the writing of this book, I also frequently remembered my deceased father, Shaukat Ali, who always challenged me to be panoramic in my scholarship despite the pressures from academe to be a reductionist "expert."

The most important acknowledgment will perhaps be to those who read this book with the same breadth of interest and help to continue the conversation on grappling with the challenges of human consumption, the environment, and development.

INTRODUCTION: ALCHEMY OF A MATERIAL WORLD

Arise order,
out of necessity!
Mock, you crystals,
with all appearance of chiseled design,
our hope of a Grand Artificer.
—John Updike, "Ode to Crystallization," 1985

The highest permanent settlement in the United States is a rather elemental town whose very name suggests its material past. One might wonder why humans ventured to the lofty altitude of 10,150 feet, braving storms, nosebleeds, and respiratory stress. The answer, in a phrase: mineral wealth. Leadville, Colorado, beckoned prospectors in search of gold, silver, and lead after 1878, when some itinerant diggers in the Rockies struck a persistent vein of the lustrous shiny mineral known as cerussite. This was a troublesome rock for the prospectors—it covered a potentially gold-bearing deposit—but these hardy souls wanted to extract the best from even the worst of materials. Although they did not know the complex chemistry behind the deposits they'd encountered, the prospectors had the advantage of experience in this geological puzzle. They knew that silver was often found close to lead ores of this kind. When they traced the vein, it did in fact lead to a higher-grade ore deposit of *argentum*—the Latin name for silver.[1] Within days of this discovery, the area was teeming with men in search of a fortune in silver, which at one time was synonymous with money, just as gold is today. (As a proud resident of Leadville informed me during a recent visit to the settlement: "Hey prof, those French folk still call their money *l'argent*"—which quite literally means silver—"so we ain't no second fiddle to gold!")

Human ingenuity and enterprise came to work in Leadville, as entrepreneurs of all professions gathered to share the wealth beyond the mineral realm. A huge tent was brought forth from the Centennial Exhibition in the East to provide sleeping quarters for the prospectors and miners. More permanent housing and businesses were quick to follow the mineral mania. *Scribner's Monthly* in October 1879 described an establishment in Leadville known as the "Mammoth Sleeping Palace": "Among the first in Leadville there happened to be a merchant who . . . built a vast shed of slabs and filled it with rows of bunks two tiers high capable of accommodating 500 sleepers. . . . He furnished a bed for fifty cents, and posted his rules: No talking or laughing, or singing, or drinking." The austerity of life in such areas was tempered by the promise of fortune. Each miner was motivated by that spark of finding just the right nugget of metal.[2]

Silver became the metal of choice for coinage in 1890 under the Sherman Silver Purchasing Act. Yet, as history has shown, humanity has faltered time and again at the behest of fashion or utility about its wants and needs for minerals. One year silver was in vogue and the next year it was not. The miners in Leadville could celebrate their fortune only briefly, since the same act was repealed by Congress in 1893. Preferences had again shifted toward gold.

Although the demand for one material, such as lead or silver, may be replaced by another, like gold or platinum, the story behind our quest for materials remains the same. We live in a world where attempts at building human order out of random natural chaos require us to manipulate materials in myriad ways. Often we try to extract a valued material from a lump of rock that might otherwise appear to be quite useless, as did the miners in Leadville. At other times we mix materials to increase their strength, malleability, or sensual appeal. Lead might be alloyed with tin, for example, to form pewter, whose ornaments continue to be favored by collectors, despite the stigma of lead toxicity.

At their core, the chemical games that we play with natural elements have fascinated and eluded us since time immemorial. Inspired perhaps by the simplicity of three primary colors, red, blue, and yellow, intricately giving way to an infinite array of shades, the ancient Greeks thought that there were four primary elements (*stoichea*). To each of these "elements" they ascribed a crystalline shape—reinforcing order by way of symmetric geometry. In his dialogue *Timaeus*, in 360 B.C., Plato ascribed the following shapes to the four elements: earth (cube), water (icosahedron), air (octahedron), and fire (tetrahedron). An even earlier Greek philosopher, Democritus, had suggested around 450 B.C. that all material consisted of fundamental particles, which he called atoms—the Hellenic word for "irreducible." However, over the next two thousand years, the further disaggregation of the four key elements or the configuration of atoms bedeviled alchemists in search of material transformation. Some, like the eighth-century Arab alchemist Geber bin Hayyan (whose name is linked to the etymology of the word "gibberish"), were motivated in this quest by a search for life-giving medicines and magical potions. Others, like the thirteenth-century clergyman Albertus Magnus, were interested in generating wealth from dirt, as many fables of turning lead to gold suggested was possible.[3]

The ultimate dream of the alchemist was to find that perfect rock that could provide the transmuting power of conversion of materials from lead to gold and also be a panacea for curing human physical ailments. This mythical rock, known in the ancient world as *lapis philosophorum*, has persisted in human imagination from ancient times to this very day. Lest we forget the power of this quest in popular culture, even Harry Potter's success started with a volume titled after this enigmatic stone!

Despite numerous expositions and experiments, the true nature of the building blocks of all materials eluded scientists for centuries. The power of myth spurred them on—from rock to rock and lab to lab—in this quest. It was not until 1661 that the aristocratic Anglo-Irish chemist Robert Boyle suggested that there were more elements—perhaps a dozen or so in his view. A hundred years later, the French chemist Antoine Lavoisier published his landmark treatise *Elements of Chemistry*, in which he listed thirty-three elements. However, he too was quite off the mark in his list, which included physical phenomena such as light and heat as material "elements."

In a remarkable convergence of genius, the modern material code was first structurally deciphered and tabulated in 1864, twice, by two scientists working entirely independently—the Russian chemist Dmitri Mendeleev and the German chemist Lothar Meyer. Both constructed a very similar table of elements that would be the precursor of the chart that graces every contemporary chemistry classroom. The unifying theme in the table was that everything was composed of structurally similar atoms. Later, through the work of Henry Moseley in the first decade of the twentieth century, people learned that the atom had a nucleus with a positive charge, and it was this charge that defined each element. Particles that we now call protons conferred this charge and were the essential differentiating characteristic between elements. Other particles in the atom might change reactivity and physical properties of the material, but the element would always remain essentially the same as long as it had the same number of those imperative positive particles.[4]

The periodic table of elements is our most indelible guide to the science of all stuff, and thus also a fundamental constraint over how we construct materials to meet our needs and wants. There are ninety-two naturally occurring elements in the periodic table, of which seventy are metals. Our environmental ambitions are confined by how the elements in this table interact with one another to form those rudimentary bonds manifesting themselves in minerals. It is with these minerals that we can dream of fueling our future. Whether constructing a dam, a skyscraper, a solar panel, or a silicon wafer, we must contend with minerals from the earth and their constituent elements. Even a humble garden or a productive pool of water lilies requires minerals to be absorbed for growth. In the case of vegetal mineral uptake, however, it is easier to return these elements to their natural state than it is for those elements used to form more permanent infrastructure and machinery. Human beings require certain minerals, such as calcium, magnesium, potassium, and sodium, on a daily basis for proper metabolic functioning, while others like lead and mercury are poisonous.

Regardless of our own biological needs, we are in an era of unparalleled consumption of all minerals. How we use our mineral resources, and whether they eventually can be recycled back to their original state, is a consideration of increasing importance. No matter how optimistic we may be, certain resources are destined for depletion, given that they are locked into built infrastructure or their chemical conversion back to original states renders their recovery economically unviable.[5]

When mining started in Leadville in the mid-nineteenth century, science had progressed far enough that many of the alchemists' conjectures about converting lead to gold had been put to rest. We knew by then that basic chemical reactions could not convert one element into another, but only combine or divide existing bonds between elements. Lead could naturally combine with oxygen to form lead oxide, which would be physically very different from the metal, but still elementally composed of lead. Yet there was another surprise in store for science. In 1898, a bright and spirited young Polish-French chemist, Marie Sklodowska-Curie, her husband, Pierre Curie, and their colleague at the Sorbonne, Henri Becquerel, discovered that it was indeed possible to convert one element into another through changes in its nucleus. The British chemist Ernest Rutherford also contributed significantly to understanding the dynamics of this phenomenon.

The euphoria that one might expect from a finding of this magnitude was not initially felt by these great minds. The Curies, Becquerel, and Rutherford were motivated not by material gain, as many of the alchemists had been, but rather by a puritanical love of science. As Madame Curie's daughter and biographer Eve Curie recounts: "The 'moment of discovery' does not always exist—the scientist's work is too tenuous, too divided, for the certainty of success to crackle out suddenly in the midst of laborious toil like a stroke of lightning. Marie, standing in front of her apparatus, perhaps never experienced the sudden intoxication of triumph."[6]

Yet the most remarkable aspect of this discovery was that a natural process, which Madame Curie called radioactivity, spontaneously caused the elemental change to occur. The alchemist's dream of converting one element to another was thus not only scientifically possible but empirically observable in nature. Indeed much of our world's energy depended on it, since it is a nuclear reaction in the sun that allows life to flourish on Earth. Our material world would never be the same: with this essential discovery about elemental transformation, the nuclear age had dawned.

Furthermore, chemistry and physics, as well as the social sciences and humanities—all of which had diverged as separate fields a few centuries earlier—began a gradual convergence toward a unity of knowledge, which the evolutionary biologist E. O. Wilson in 1999 called "consilience." Madame Curie became the first person to be awarded a Nobel Prize in physics as well as in chemistry. Natural scientists also became increasingly involved in political matters, including such luminaries as Albert Einstein and Linus Pauling, who became active in discussions about nuclear warfare. In 1962, Pauling became the first Nobel Prize–winning scientist to also be awarded the Nobel Peace Prize for his work on nuclear disarmament.

The opening of the twentieth century marked troubled times for the world. Our growing knowledge about the material world allowed us to turn it against ourselves. Wars were waged, and all sorts of materials were exploited for most unsavory tasks—particularly the manufacturing of weapons. Mining boomed across the world, especially of metals like lead and copper, used to produce the arsenal needed to fight the battles taking place on land, in the air, and at sea. After World War II ended in 1945, the consumption of materials for arsenals was kept aloft by the Cold War, and metals like chromium saw an astronomical rise in demand owing largely to their specialized weaponry uses.[7]

This consumption surge was not confined to the military-industrial complex; it also included individual consumers. Following World War II, the world's population rose threefold in five decades, and so did the wants and needs of each of these individual consumers. This increase was most dramatic in the United States, which was then and continues to be the most dominant consumer society (fig. 1).

Prosperity in the postwar years also increased the intensity of consumption per person, and there were few incentives to save or conserve. As we discovered more materials and ways to construct new products, our impulse for seeking yet further treasures carried us forward to new frontiers of consumerism. These materials needed to come from somewhere on the planet, and the race to find resources acquired a feverish pitch for individuals and corporations alike. Yet unlike the gold rushes of yesteryear, we began to consider the value being added to products along far more complex supply chains.

With the advent of plastics in all their many forms in the twentieth century, molecules also began to form much longer chains in our material world. Unlike most earlier materials that had relied on metals in some form or another, plastics were based on carbon—the element with the most abundant permutations of compounds in the universe. Carbon had unobtrusively lurked in the periodic table as a chameleonic character that

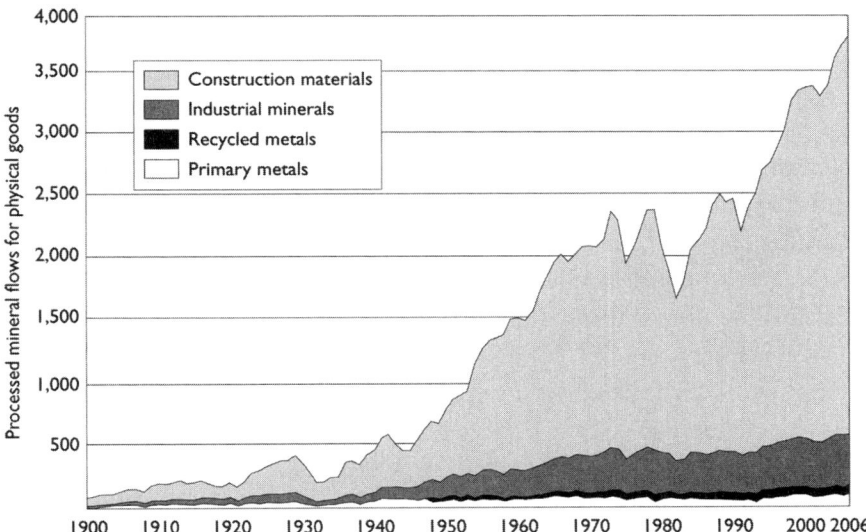

FIGURE 1. Raw nonfuel materials consumption growth in the United States. Figures are in millions of tons. (U.S. Geological Survey)

could be seen in nature as coal, graphite, or diamond. Its ability to form complex bonds and chains with other carbon atoms and various elements was well known to scientists as essential to most molecules of living organisms—giving rise to the entire field of organic chemistry, which deals solely with the compounds of carbon. But the needs of the consumption frenzy that came in the middle of the twentieth century required chemists to consider carbon compounds with more vigor.

Powered by that prolific former of carbon chains, petroleum, the chemical industry was able to create a panoply of products that could be consumed cheaply and had negligible metallic content. However, to build the infrastructure necessary to produce and transport products—even these miraculous plastics themselves—we still needed base metals. Despite the ability to stretch and link renewable molecules of organic compounds, we remained dependent on our basic earthly goods—no matter how far removed our plastic bottle seems from anything found in nature.

Soon we began to realize that there were limits to what we could do with extractable materials from the earth's crust. More energy was needed to consider the power of living organisms as a renewable resource. We could not grow a tree of silver or copper, but perhaps we could grow specialized vegetal materials that might produce the same material characteristics as the prized metal, or find some fruitful combinations of organic and inorganic constituents. Just as our cereal boxes tell us the nutritional value of organic compounds like sugars and carbohydrates to provide us with energy, so too do they mention the much smaller but equally essential quantities of some metals, such as zinc and iron, that keep us functional. Just as nourishing the individual human requires a complex assemblage of elements, so does the sustenance of an entire society of sophisticated human consumers.

In downtown Leadville is the National Mining Museum, an impressive red-stone building with a vestigial air of nineteenth-century grandeur. Hanging by its entrance is a telling sign that reads: "More minerals have been consumed since the start of World War II than in all of the rest of human history." At the beginning of the twentieth century, roughly twenty elements of the periodic table were exploited for commercial product manufacturing. Today, manufacturing draws from all ninety-two of the naturally occurring elements. The complexity of our material usage has thus gone up dramatically.[8] We now manufacture plastics like nylon and Teflon, and carbon composites like those used in most modern commercial aircraft, which have intricate production processes involving hundreds of material inputs. These materials are particularly persistent in the environment and cannot decay easily like most other earthly materials that find their way back to their elemental origins. Often the oceans become the repository of persistent plastics, as exemplified in their high concentration in the North Pacific subtropical gyre between Hawaii and the California coast. By one estimate, man-made materials now account for six times the naturally occurring zooplankton floating on the surface of this region.[9]

Even though we have become more efficient in using materials and making our products perform better with fewer inputs, overall consumption has continued to rise dramatically. According to a study by the U.S. Geological Survey, from 1975 to 1995 total consumption of materials swelled by 67 percent; what is more, the amounts used, by weight, have moved from 50 percent renewable in 1900 to only 8 percent renewable now.[10] For some, this is clearly a sobering statement, as it shows the immense impact we have had on our environment.

During a visit to a mine in Australia several years ago, I saw a young child of a miner stare in amazement at the huge expanse of the open-pit mine and remark spontaneously: "What did you do to this place?" For the child's father, the gargantuan pit and its enormous trucks carrying ore to and fro made an exhilarating statement of a different sort, showing how industrious human beings have been in developing modern amenities. Certainly, one cannot overlook the progress made with our material usage. From the most basic advances in petroleum-based plastics used for artificial heart valves to the discovery of the metal niobium as a capacitor that enables cellular phones to function, we have harnessed our material wealth for development. Yet we cannot ignore that, for every useful product, we also create another product of much less utility. Even if those of marginal utility exist merely to meet human wants, rather than needs, we can still consider more environmentally considerate means of producing these materials.

Regardless of whether you think consumption should be prioritized by needs or by the ecological impact of production processes, there is little doubt that rising human consumption of materials is our most seminal environmental challenge. However, concerns about resource depletion should lead not to despair but rather to solutions and effective planning. In our quest for sustainability we are faced with two choices. The first is to try to ensure that all natural resources are maintained at adequate levels to provide for an indefinite supply (also known as the "strong sustainability argument"). Essentially, we could seek to reap only as much as we can grow. In this case, we would try our utmost to focus on renewable resources that can be created on human timescales—essentially vegetal and animal materials—and try to harness whatever we can from those materials. Thus wooden materials and vegetal plastics would be preferable to metals and petroleum-based plastics. Consider polylactic acid, a molecule made from the humble potato that is now being developed as a vegetal replacement for plastic in cups, soup containers, and food packaging. Even the colossus of all consumer merchandise, Wal-Mart, has taken the plunge and ordered multimillion pounds of this product.

Despite such promising examples, the problem remains that there are no substitutes for nonrenewable resources in many of their uses. At present, we cannot build a dam or skyscraper solely out of wood and vegetal materials, although research on natural composite materials continues at various levels. More than a decade ago, I visited the Sikorsky Aircraft plant in Connecticut to see how composites were being used to develop the next generation of helicopters. The engineers at the time were most apprehensive about the efficacy of these materials, and as a student of industrial ecology I was skeptical of the comparative difference in environmental impact between conventional and composite materials. Now, however, we have the data to show that such materials are durable and environmentally more promising than their metallic predecessors. Even if we squeeze the last ounce of effective molecules from potatoes, corn, and other agricultural products, though, we may still be left with some residual uses that require nonrenewable materials. Here we need to consider another strategy.

This second option argues in favor of focusing not on the resource quantity itself but rather on the aggregate stock of natural and human capital required to harness the material from its dispersed state in all our consumer products—also known as the "weak sustainability argument." Let us not dismiss this second option simply because it is labeled "weak." Granted that many proponents of this view have erroneously suggested that well-being can be equated with income, since they view most natural capital through the lens of market goods. However, the essential assumption in the efficacy of weak sustainability rests on the ability of humans to increase efficiency, develop new technologies, and—most significantly—change the process of material flows in our world.

Consider the silver mined in Leadville more than a century ago—likely the silver was sold on the commodities market, passing from merchant hand to manufacturing hand, where it was converted into coins or tea kettles or perhaps even photographic processing solutions. Where is that Leadville silver today? We have no idea. We don't know how it could be retrieved from the world of entropy. This example is particularly unfortunate, for, as a relatively noncorrodable metal, silver's elemental nature makes it eminently recyclable. But in order for such metals as silver, gold, and aluminum, as well as other minerals, to be reused and not lost forever, we must invest enough energy to bring them back from their wasted use. Human capital is required to make that happen, and this capital grows with every passing day—both in quantity and quality. However, it involves a great deal of foresight. We have undeniably made the task of retrieving these precious materials much more difficult because of poorly designed products. If the users of silver, such as photographic manufacturers, had considered the potential for retrieving the metal after its use, they might have designed products and processes very differently. They may have looked for a material that would be most easily retrievable and less likely to be chemically converted to more intractable forms over time.

Thus, according to the weak sustainability premise, we need a major conceptual shift about how our economy functions. Using the analogy of water-based systems, ecologists have compared the current system to a river where materials generally flow in one direction. What is needed instead is a "lake economy," in which stocks of materials might circulate indefinitely, so long as the inputs and outputs are effectively managed. The good news, as you will discover in the pages of this book, is that such changes in paradigms about our world are indeed possible without dismissing consumerism. For just one example, we can design a computer's consumer cycle to be modular. When we want to upgrade our computer, we can simply replace the outmoded chip or screen or hard drive, requiring manufacturers to recover the materials from the old modules, instead of exchanging a whole machine for a new one incorporating newer technology.

Effective governance systems are crucial to make these changes occur, and there are countless examples of poor governance making problems much worse. In the case of computer recycling, the lack of regulations on the transport of computer waste to Asia has created perverse incentives for manufacturers to trash old equipment rather than recovering valuable components for reuse, or, better yet, designing more modular products. However, in some places, we are already beginning to see signs of progress through effective governance structures. For example, landfill taxes in Denmark increased the reuse of construction debris from 12 percent to 82 percent within a year of implementation in the late 1990s.[11] While many might consider such actions to be "low-hanging fruits," there are still enough branches dangling before us if we only looked up and tried to reach them. Once all the fruit is plucked, there still may be other options to consider, if we save the seeds and know how to plant them.

The people of Leadville seemed to still know how they might plant these seeds of agency in even the most harsh and unrelenting environment.

Almost a hundred years ago, the miners endured a downturn in their fortunes when silver prices plummeted. The community rose to the challenge to try something totally new. Knowing that they had very few resources at their altitude and terrain, the residents of Leadville brought their ingenuity to bear on that most abundant of all solid natural features of cold Colorado winters—ice! In the winter of 1896, more than 250 men from the community banded together to build a huge structure of ice and wood. The "Crystal Palace" was to attract visitors to their desolate abode. Within thirty-five days, the ice palace was completed—the largest of its kind ever built in world history. It covered more than 58,000 square feet, utilizing 180,000 board feet of lumber and 5,000 tons of ice. Its towers reached 90 feet high by 40 feet wide, and the palace encompassed five acres of ground. More than 250,000 visitors came to Leadville to see how this little mining town had transformed its primary source of livelihood. Even if just for a fleeting winter season that brought not much financial gain, the spectacle left posterity with a momentous memory of human perseverance.

Amid the golden fall of aspen foliage, I visited the old opera house in the center of Leadville to attend a meeting organized by the community in 2006. On the faded walls of the building was a flashy banner that heralded the theme of the event: "Too Tough to Die." Though many towns worldwide have ebbed and flowed with material fortunes, this little community seemed resolute to survive. The population of the city has diminished from over forty thousand in its heyday to less than three thousand, but the Leadville community that I encountered at the conference was ready to reinvent itself. The ice palace was still a beacon of strength for them as they pondered a transition to become a tourist destination and an educational base for the local community college. There were even prospects that new technology and rising metal prices would allow for the old molybdenum mine nearby to reopen. These various trajectories may or may not work for Leadville. Indeed, many other mineral communities would not even have such options to consider.

Yet the proclamation of defiance against forecasted doom had been made, and delegates from many other resource-dependent communities seemed to follow the call at the event. Despite the idealistic tone of the proclamations, the significance of the event was more profound than any little town could possibly fathom. There was a conviction to invent a future of prosperity. Through their history, and this sign, the people of Leadville seemed to almost instinctively understand the human impulse to consume minerals, and the positive and negative stakes wrapped up in this consumption. Their town serves as a microcosm for the world at large: the peril and promise of extraction, the inexorable human demand not just for progress but for possession of riches, and the need to be perceptive about how we consume in the future so as not to destroy our environment. The lessons that we might glean from understanding our intimate relationship to materials are immense and crucial for human survival at a planetary scale. We are just beginning to embark upon this exciting quest.

PART

I

THE PLEASURE OF TREASURE

When Diamonds are a Legend,
And Diadems—a Tale—
I Brooch and Earrings for Myself,
Do sow, and Raise for sale—

And tho' I'm scarce accounted,
My Art, a Summer Day—had Patrons—
Once—it was a Queen—
And once—a Butterfly—

—Emily Dickinson

1

WHAT LIES BENEATH: THE MATERIAL AND MYSTICAL ORIGINS OF MINERAL WEALTH

> For whatever is his own is well concealed from the owner, and of all the treasures, it is our own that we dig up last.
> —Friedrich Nietzsche, "On the Spirit of Gravity," *Thus Spoke Zarathustra*

The grand cemetery in Medina, Saudi Arabia, is a sight to behold, not because of any ostentatious splendor that one might expect, but rather for the vast expanse of dusty mounds and nameless graves that stretch out before all visitors. For centuries, kings and paupers alike have been buried here without any fanfare, as a testament to our common and very earthly origins and destiny. The Wahhabi sect of Islam is known for its more sinister fundamentalist traditions, but there is a benign simplicity in its insistence that human remains be sent back to their elemental origins with minimalist zeal. During my first visit to this stark and somber place, usually teeming with thousands of visitors, I pondered on the diversity of ways our species deals with death. Some human societies built elaborate tombs and even buried their dead with treasures for the afterlife, like the Egyptians; others opted for the ultimate annihilation of the body through cremation. The Zoroastrians, whom the German nihilist philosopher Friedrich Nietzsche so admired, disposed of their dead on elevated structures known as the towers of silence. Here, vultures and other avian carnivores were allowed to feast on the carcasses, and then the bones were collected and buried. The Zoroastrians observed this practice as a mark of respect for natural cycles of life and death. Burying or burning fleshed corpses was considered a violation of earthly processes, whereas bones were a permanent earthly material and hence could be returned to the soil.

Whatever the method of mortal disposal, the basic premise remains the same: our physical beings are recycled back into the earth's crust from which we originally came. From birth to death, each human metabolizes and releases matter and energy in various forms and to highly varying degrees, depending on where they are born and what they accomplish in that lifespan. The earliest human societies seem to have appreciated that there is no escaping this reality.[1]

Apart from their figurative importance, minerals were an essential substrate for life to evolve on Earth. Long before the first cellular organism inhabited the planet, our world cooled from its gaseous and molten origins to form minerals. The unique characteristics of elemental arrangements in minerals may have allowed for life to begin. Natural scientists explain the monumental mystery of how order arises out of a chaotic world by the principle of emergence.[2] Through a system of inherent hierarchies within natural systems, seemingly random occurrences take an ordered form in synchrony over time. For example, waves that repeatedly move sand back and forth form a rippled pattern over years, or birds responding to certain meteorological cues independently exhibit flocking behavior. In order for chemicals to organize and propagate a cell, some organizational substrate may have been essential. In the view of biochemist James Ferris, certain clay minerals might have provided the requisite substrate for life-giving molecules to be synthesized on their own. Biotic emergence could thus have been catalyzed by the presence of minerals. Through experimentation with a volcanic mineral called montmorillonite, which is often mined for use in filters, Ferris has been able to synthesize such biologically significant molecules as ribonucleic acid (RNA) from much simpler compounds.[3] This marvelous mineral was most likely present around the time that cellular life emerged on Earth, around 4 billion years ago, and may have been the lab space for life to find its way.

Once life evolved, minerals were also a source of nourishment for the earliest organisms. Several early organisms evolved in the absence of light and used minerals, particularly iron, as their source of energy. Highly specialized bacteria that love heat and avoid light were well suited for the anaerobic environment of ancient earth. These "hyperthermophilic" bacteria feasted on iron and may provide us with greater insights about the evolution of our own blood, which is colored by iron.[4] Minerals are thus integral to the metabolism of arcane microbes as well as precocious organisms, such as us.

Our history as a civilization is eventually taken over by the earth. Layers of dust and debris bury us; no matter where we are or how we bury ourselves, eventually the only way to uncover our past is to dig deep. What lies beneath our feet usually tells us about some bygone age, our ancestors, and their stories of survival. No wonder humans are so fascinated with excavations; as we unearth history, we inevitably find some objects of value associated with human settlements. Even if it is a flint spear or stone club, implements are usually found in the ruins.

Yet the impulse to collect such objects, cherish them in our lives, and then pass them on to our progeny has also served us well as a species. Although other species engage in collecting objects for various utilitarian ends, such as building a nest (most birds), constructing a dam (beavers), or storing food for winter (numerous mammals), none have taken the next step of aggrandizing them for pleasure and utility the way humans have. Furthermore, most other advanced organisms, such as chimpanzees, use the material they collect for tools, such as stones to crack nuts, only near the source of the implement. According to one study, 90 percent of all implements used by chimpanzees at a nut tree come from within about two hundred yards of the tree. The ability to explore, plan, and carry the tool over a longer distance to serve its purpose is apparently a distinctly human trait.[5] Collecting and using material in the persistent way of humans may have played an important role in our rise to prominence along evolutionary pathways. Recent scholarship in evolutionary theory reveals that the impulse to consume and seek out new materials, particularly beneath the surface of the earth, may have given us distinct advantages against the elements and other organisms.[6]

Yet we must also contend with the probability that excessive consumption and depletion could have led to the demise of numerous civilizations. The societies of Easter Island, the Mayas, and the Aztecs, for example, may have been destroyed when they deforested the landscape beyond its ability to replenish minerals in the soil, or perhaps due to other factors beyond their control.[7] The extraordinary interdisciplinary scientist Jared Diamond reminded us of this prospect in his best-seller *Collapse*, which in turn was challenged by archaeologists and anthropologists as too linear an explanation for societal demise. In the fall of 2007, a group of such researchers got together in the offices of the Amerind Foundation in Tucson, Arizona, to raise objections about Diamond's compelling hypotheses regarding ecocidal ruin of societies. Their main contention was that human societies have more agency than environmental determinism may give them credit for. They interact with their environment in dialectical ways that can usually not be explained by a grand theory. Humans also have the ability to learn from mistakes in ways that concerned environmentalists might be reluctant to consider, lest we sink into complacence.[8]

Regardless of the debate over unified theories of civilizational collapse, finding the balance between digging deep for resources and respecting the fragility of our earthly resource base has been our most seminal challenge as a species. We have often swayed from one extreme of the consumption spectrum to another in our preferences for individual consumption, without fully understanding the complex implications of the choices that confront us. We sometimes chastise ourselves for wanting more than we need, and for our focus on embellishments rather than essentials. At other times we try to find ways to preserve cultures whose most essential characteristics might be manifest in seemingly redundant works of artistry. The belief that humanity has moved on a trajectory from need-based life to want-based consumerism is widespread. Yet, in even the most subsistence-oriented societies, we find a vibrant desire to adorn and imbibe the superfluous and unnecessary.

To support such desires for material adornment, even the earliest cultures needed mines. The oldest known mining operation so far discovered can be traced to the mountains of the tiny African country of Swaziland. Nestled atop a ridge named Bomvu (which means red in Zulu) lies a cave that might initially seem like any other natural cavity. However, on closer inspection, this marvelous cavern has all the signs of human intervention. Chiseled arduously by our ancestors, the Lion's Cave, as it is now called, has an attractive gray crystalline coating of an iron mineral called specularite. Deeper into the cavern there is a large boulder, behind which hides a mineral "gallery." A radiocarbon date of the site places its designs at 43,000 years before the present, during the period when only stones were used as implements. So far back in time, the lives of those who dwelled here must have been rudimentary and difficult. Yet the people of this culture were mining not some essential material for survival but rather an indispensable pigment for rituals: ochre. The miners of Lion's Cave sought this iron-rich pigment to use its colorful hues of rust and yellow for facial cosmetics and for marking tools and implements.[9]

Such examples of the use of ochre can be found from more incidental occurrences of art on rocks (rather than from mined activities) even earlier on. For example, the French Riviera's most celebrated archaeological site, Terra Amata, has evidence of the use of ochre going back more than 300,000 years. More vibrant and identifiable uses of ochre can also be found in the picturesque Becov region of the Czech Republic (200,000 years before the present) and on aboriginal rock art in Australia (75,000 years ago). Thus the first known mine was not for physical necessity, but for cultural connectivity through religious adornment or artistic expression. The anthropologist Cyril Stanley Smith once observed that "aesthetic curiosity, not necessity, was the mother of invention," since it was the decorative—not utilitarian— value of metals and minerals that led humans to further explore their uses.[10] The paths to human ingenuity are paved as much by wants as by needs.

What is also striking is the association of class structures with mineral wealth accumulation in early societies. In March 2008, the *Proceedings of the National Academy of Sciences* reported the discovery of an ancient gold necklace on the shores of Lake Titicaca in Peru dating to as much as 4,000 years ago, making it the oldest gold artifact recovered from the Americas in a region often associated with hunter-gatherer societies at the time. The authors wrote: "The presence of gold in a society of low-level food producers undergoing social and economic transformations coincident with the onset of sedentary life is an indicator of possible early social inequality and aggrandizing behavior and further shows that hereditary elites and a societal capacity to create significant agricultural surpluses are not requisite for the emergence of metalworking traditions."[11] This discovery showed that, for better or for worse, our treasure impulse was perhaps more pervasive in even the earliest societies than previously thought.

Flint Stones to Metals

Some of the oldest need-based minerals were brought from the earth's interior to its crust by volcanoes. The glassy congealed form of lava that gives us obsidian, a sharp cutting stone, was perhaps the first material to be extracted by our ancestors for utilitarian purposes. We find the first signs of obsidian mining on the slopes of Mount Ararat, a majestic volcano in present-day Turkey where Noah is supposed to have landed the ark after the Great Deluge. As the lava from the volcanoes in this region rapidly cooled, it turned into sharp black deposits of obsidian that made ideal material for cutting implements.

The utility of obsidian was similar to that of another telltale rock that signals human activity for archaeologists—flint (or silex). Use of flint can be traced as far back as 2 million years before the present, to humanoid creatures who predated modern humans. Flint tools were widely dispersed, since they were fashioned largely from collected stones over wide expanses of territory. In contrast, obsidian is concentrated around volcanic areas, and hence led to a more concentrated extraction regime. Its sharpness is valued even in modern tools: today this mineral is used even in implements for cardiac surgery.

Moving west and south from the plains of Ararat, toward the center of the Anatolian peninsula, we come across a remarkable set of ruins that constitute the first known urban area ever built. This place is currently known as Çatal Höyük in Turkish and is considered one of the most active and promising archaeological sites in the world for Neolithic activity. One might wonder why people in Neolithic times gathered to form a settlement or city in this remote area. Its location—bordering the edge of a rich alluvial plain—made it a choice setting for food production. But it is likely that there was an even stronger reason. The city was formed at the crossroads of trade routes, and its proximity to obsidian deposits at Lake Van, Aksaray, and Bingol suggests that this community was likely nurtured by extractive resources. More than six thousand people lived here, and they created the earliest form of pottery firing and other sophisticated visual arts with pigments on primitive textiles. Some of the paintings show erupting volcanoes in the background—the wellspring of the town's mineral wealth. The obsidian was traded for seashells brought from coastal communities, which were another primordial source of material fascination for our ancestors. Shells, volcanic glass, and raw stones were our earliest coveted treasures.[12]

The survival of civilization depended on exploitation of minerals in their raw, stony form—the central importance of this fact is reflected in the names we give the Stone Age and Neolithic periods of human history. The ability to harness elements, specifically metals, from specific kinds of stones, known as ores, was a seminal step in human development that took place around 2500 B.C. Anyone who has watched *The Flintstones* will appreciate the infrastructural strides humans have made since moving from stones and wood to metals. Yet we mustn't underestimate the permanence of stone as a material in society, both then and now. Stones and aggregate are necessary for any elaborate construction. Even now, as we contemplate the potential demise of our civilization as portrayed in bestselling books like Alan Weisman's modern classic *The World Without Us,* the permanence of stone structures outstrips that of modern concrete and steel. Stone construction, along with persistent plastics, might be our most ubiquitous and lasting legacy.

Metallic material complemented stone in building humanity's civilizational infrastructure. The move from Neolithic culture to a society able to extract and use metals was a giant leap for humankind. It is no accident of language that we have named periods in human history after important metals: the Iron Age, the Bronze Age, and various permutations of Golden Ages.[13] These metals have influenced culture and development in ways that we are only beginning to understand. But metals vary tremendously in their properties of strength, durability, and abundance. Metals like aluminum were once utterly abundant in the crust of the earth, but they were so intimately bonded with other elements in rocks that it was impossible for our early ancestors to extract them. Others, like copper, were found in metallic form as huge, readily usable nuggets and curtains of native metal.

Most metals are elements and hence quite fundamental in their chemical characteristics. Perhaps our ancestors subconsciously understood this property, leading them to use metals for utility and adornment alike. Appreciating the value of fire, as the paradoxical natural force that could both hurt and heal, was an essential milestone in human history, dating back tens of thousands of years. Initial experimentation with heating rocks while cooking food may have led to the development of the first smelting furnaces. Smelting is the process of heating ore to unleash metallic elements. Copper was the earliest of the metals to be appropriated for its utility, as it was initially found quite abundantly near the earth's surface, and can be extracted by heating certain kinds of rocks at fairly low temperatures. Since copper was rarely found in a pure state, during smelting it often combined with other metals included in the ore. Metalsmiths later became more sophisticated at calibrating the ratios of rocks to create the desired metallic state. Metallurgists, as they came to be known, blended copper with tin and arsenic, creating the vital alloy that defines the earliest metal age: bronze.

The first examples of Bronze Age artifacts can be traced to around 4000 B.C., from the area consisting of modern-day Iraq and Iran, although the techniques may well have been discovered independently in East Asia and Africa. The advent of the Bronze Age in the Americas didn't come until much later, owing largely to the deposits of copper being found at higher, less accessible altitudes. In the Mediterranean, the most organized and concentrated arrangement of copper smelting developed on the island of Cyprus. Most significant, the Cypriot copper galvanized a new era of trade among various parts of the region. Copper from this island was shipped off to Egypt, Sardinia, and Anatolia and used for fashioning sculptures of gods and goddesses as well as tools and utensils. The impact of this tiny island's treasure load of copper during the Late Bronze Age is phenomenal. For example, excavations around the town of Phorades show that metalworkers in the region had constructed an artificial embankment around their workshops, where more than six thousand fragments of rims, walls, and bases belonging to freestanding cylindrical furnaces were discovered. To this day, one can find large heaps of gray slag in the Sia Valley left over from those distant days of copper smelting. As the archaeologist F. L. Koucky noted: "It is unfortunate that the slag heaps in Cyprus have not been treated as important antiquities, for they record an effort greater than the building of the pyramids or the construction of many great cities of the past."[14]

Although iron and aluminum were far more abundant in the earth's crust than copper or tin, the limiting factor in the development of iron smelting was—again—fire. Getting the flame to the temperatures needed to separate iron from its ore required high-quality fuel and the ability to create specific conditions of airflow in the furnace. Moreover, unlike copper, which flowed forth nicely after smelting from its ore, iron melted and congealed into a spongy mass or coalesced into a circular "bloom" that was interspersed with pieces of waste material from the smelter, known as slag.

Before the discovery of iron-smelting techniques, the purest naturally available iron came from meteorites, whose perilous journey through the cosmos eroded and consumed all vestiges of other materials except for the most resilient metals. Celestial connections with metals were thus made by the earliest miners without their knowledge. The flaming attrition of these rocks in our atmosphere acted as a natural smelter. With time and experimentation of various kinds, metalsmiths perfected the harnessing of iron. To smelt iron in optimum amounts, heat must be maintained for a longer period of time than for copper, but still at a temperature lower than the melting point of the metal. This technique was first discovered by the Chalybes and the Hittites in Anatolia around 1500 B.C., and these people jealously guarded their knowledge for centuries. These ancient mining centers of the Middle East became the epicenter of trade and commerce for much of the old world (fig. 2).

Professionals who refined metals were often considered an exceptional class by past societies. Metalsmiths were seen as people with almost magical powers over the earth. They often practiced endogamy and married within their professional circles. In African societies, there was a particular mystique surrounding the practice of metalwork. As the writer Ndinga-Mbo states: "In venturing into the bowels of the earth, the smith braves the order of nature. The other phases of metallurgy are equally acts that bring the smith in competition with the gods." As revered as they were in some societies, however, in others they were reviled. In certain African traditions, there is a negative ascription of power to the acts of the miner and the metalsmith. According to Youssouf Cisse, in the belief system of the Mande people of West Africa, "opening a mine, extracting a mineral, is equivalent to murdering a superior being, the earth, this mother of mothers." The act is therefore accompanied by offerings to animal protectors of the earth, such as the hyena, in many African traditions.[15]

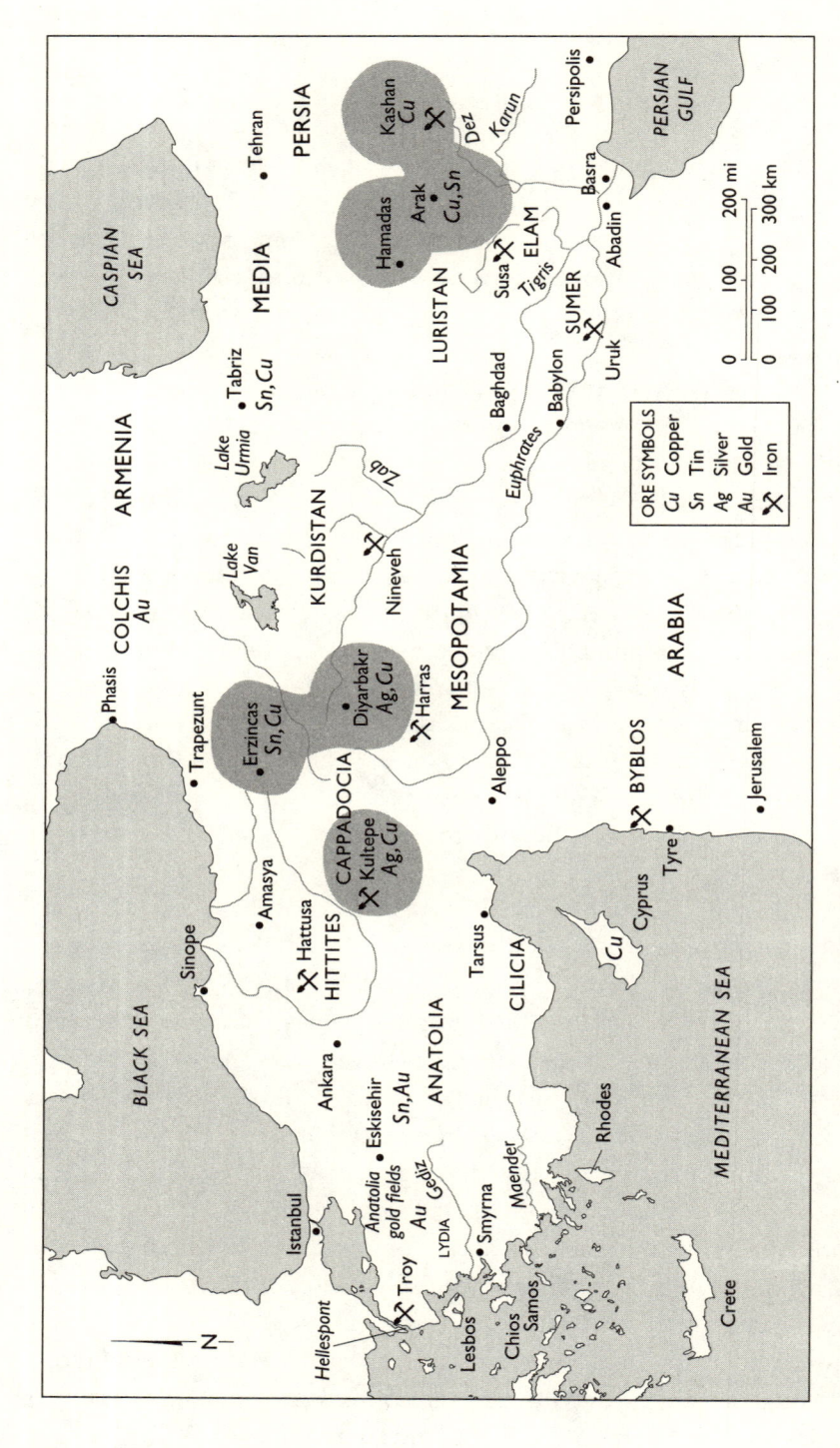

FIGURE 2. Metal mining areas in the ancient Middle East

In the Yoruba culture of Nigeria and West Africa, the deity Ogun is given the form of a blacksmith, king, farmer, circumciser, warrior, civilization bearer, and hunter. Interestingly, oral histories describe Ogun's multiple professional attributes as an indication of the paradox of urban versus rural life and the impulse to hunt. The Yoruba recognized the struggle to balance extractive enterprises—of whether to seek food or material treasures—long before oil and other riches were found near their land (in contemporary Nigeria). As Bade Ajuwon has described the life of this deity: "Ogun lived half his life in the wild and disorderly state of nature. . . . And the other half in the orderly state of human existence, but Ogun loved best his solitary life in the forest."[16]

Mining in Africa goes back more than four thousand years. Early extractive techniques to harness copper were used near Agades in modern-day Niger. African societies perfected smelting technologies to a remarkable degree without industrial advancements, defying typical assumptions that link use of metals to industrial progress. Except for one furnace developed in Burma, for example, Africa is the only place in the world where natural draft furnaces have been documented.

The belief in the potential mystical power of metals spans not just time but geographic space. Copper bracelets that continue to fascinate many new-age tourists in Arizona for ascribed healing properties were just as alluring to the Lemba people of Congo. Only priests and their spouses were allowed to wear the copper bracelets and were feared throughout the land for their power. Copper also symbolized the stages of life from birth to puberty to old age and was widely used by African societies for marking rites of passage and fertility. Iron was even more intimately linked to sexual rituals in African societies, and the image of copulation was often quite explicit in the smelting apparatus. The production of iron was allegorically linked to giving birth. In the Tshiboco furnaces of Kenya, the bellows used for air flux constituted the male component of the apparatus that rhythmically panted as air was pumped through the furnace (often described as the female torso), and a bloom of iron was eventually produced.

However, as Eugenia Herbert has noted in her detailed studies of metallic rituals, not all bellows were male. They often mirrored a male-female complementarity, and a pair of bellows themselves might be labeled "husband" and "wife." For the Mbeere of Kenya, it was essential to use skin from both sexes of goats in building the bellows to reinforce this complementarity. The right-hand bellows was always made from a male goat calf and the left-hand from a female. Natural draft furnaces in Burundi that do not require a bellows used phallic imagery in plugs that control air intake.[17]

Although furnaces were equated with sexual organs and the desire to extract copper and iron was often allegorically linked to libido in African cultures, the metal that has most fascinated humanity throughout the ages did not initially require heat for extraction. The very sight of this brilliant element was enough to enrapture most civilizations and symbolize the ultimate lust for treasure. Gold was clearly attractive because of its shiny brilliance, but also because of its durability. The metal's very stable atomic structure makes it the most chemically inert of all solid elements—it can therefore retain its shine without falling prey to the charms of oxygen molecules that bind with most metals, forming flaky salts that manifest themselves as corrosion.

The Divinity of Gold

A toothless old miner in the dusty Australian outback once told me with a provocative smile: "There is only a letter's worth of difference between God and gold." The metal has been emblematic of humanity's fascination with precious materials and our willingness to deify material value. While most religions have zealously fought against this commodification of divine attributes, the temptation to use material embellishments in halls of worship remains evident even in the scriptures. Although Moses chastised the makers of the golden calf on his way back from Mount Sinai, he also brought back precise instructions on building a glorious temple, which God ordained: "Thou shalt overlay with pure gold, within and without shalt thou overlay it, and shalt make upon it a crown of gold round it." Exodus (chapters 25–28) devotes more than eighty paragraphs to the construction of this decadent temple. The ancient Egyptians, as well as the Persians, were captivated by gemstones, particularly turquoise. Pharaoh Thutmose IV is best known for the image of him engraved next to a turquoise mine in Sinai, illustrated with sacrifices of milk and lotus blossoms to the goddess Hathor.

The apotheosis of material wealth is perhaps a reflection of humanity's desire to reward divinity with what we see as most significant. This in turn can also transform the fortunes of the clergy. In 1305 the papacy was moved from Rome to Avignon, France, much to the surprise of Christians everywhere. According to historian Roland Bainton, Pope Clement V took this step in congruence with an order from the king of France, Philip IV, banning all exports of gold. Accumulating gold inside the borders meant greater opportunities for investment and charity in the French kingdom.[18]

Later in the fourteenth century, the Black Plague destroyed a third of Europe's population. Paradoxically, this was a time of tremendous luxury and excess: the wealthy merchants and aristocrats who died of the disease left behind enormous assets in gold to their survivors. This proved to be lucrative ground for financing many institutions, including churches, that were lavished with embellishments made of gold and other precious materials.

The youngest of the Abrahamic faiths, Islam, was blossoming around the same time across southern Europe, and precious materials figured prominently in Islamic architecture and art as well. Relics made of gold have been valued by polytheistic religious creeds as well as their adversaries in iconoclastic monotheism. Even though orthodox Islam forbade men from wearing gold jewelry, no such injunctions existed for women, who in fact were encouraged to wear jewels to make themselves more attractive to their spouses. The cubic simplicity of the Kaaba in Mecca is still adorned with an enormous gold door and a calligraphic border in gold thread, symbolizing the purest Quranic inscription on what Muslims believe to be God's house.

But this indulgence is by no means exclusive to the Abrahamic faiths. Religious devotion has given us a guilt-free rendition of our material desires across cultures. Even the most self-effacing Buddhist monks have adorned multitudinous statues of their beloved Prince Siddhartha with gold and gems. In fact, the world's largest artifact made of solid gold is a magnificent Buddha statue in the Wat Traimit temple in Bangkok, Thailand. Relics of gold have been the mainstay of religious traditions because of their permanence in the eyes of believers. Deities are special because they are thought to always be there for those who need to invoke them.

Gold evokes the same constancy, chemically, in the crafting of religious relics. No wonder that in most archaeological epics, including those of modern-day fictional heroes like Indiana Jones, gold religious figures are indispensable.[19]

To many, the attraction of gold was as "natural" as evolution, and during the nineteenth-century gold rushes, the emergence of the monetary gold standard was often conveyed through images of Darwinian inevitability. A famous cartoon by Thomas Nast illustrates how gold was considered the culmination of a search for a prized currency—the "dawn of truth" breaking through the darkness (fig. 3). As the debate over paper currency was being conducted, the symbol of gold resonated with many Christian

FIGURE 3. "The Survival of the Fittest," cartoon by Thomas Nast (From David A. Wells, *Robinson Crusoe's Money* [New York: Harper and Brothers, 1876])

theologians who had come to the New World. As noted by Kathryn Morse in her study of the Klondike gold rush, "the argument about values, both monetary and moral, was rooted in the symbol of gold, and in nature, a nature created with certain purposes by a Protestant God." In contrast, those who opposed the gold standard tried to "denature" the image of gold, by showing that it was only representative of perceptions of value and did not have any inherent quality of need to humans. The critics of the gold standard represented the metal as sterile compared with a seed planted by a farmer that could add value through nourishment and growth, as labor and nutrients made it of intrinsic worth for survival. However, the primacy of gold and the rushes to find it endured even this line of critique. The great chronicler of the Klondike gold rush, Jack London, conceded in a back-of-the-envelope cost-benefit analysis that the miners had probably invested more than $220 million to build infrastructure and sustain themselves, and in turn they collected around $22 million worth of gold. Yet to him, this calculation at the turn of the twentieth century still weighed in favor of the yellow metal's worth. He noted that this monumental effort was of "inestimable benefit to the Yukon country," because "natural obstacles will be cleared away or surmounted, primitive methods abandoned, and hardship of toil and travel reduced to the smallest possible minimum."[20]

Even though gold was almost universally valued, there were some rare exceptions to the rule. In highly resource-scarce communities, such as the desert tribes of the Sahara, the mineral commodity of choice was not an inert metal like gold but a compound formed by one of the most reactive metals in the periodic table—sodium. So reactive is this light metal that it is never found alone in nature but always combined with numerous other elements, the most common of which is chlorine—forming that essential ingredient for the human palate—salt. Salt was a deity for many tribal groups, including Native Americans like the Navajo, the Zuni, and the Hopi, who all prized salt despite their many other cultural differences. Peter Bernstein describes the comparison between gold and salt in the eyes of Saharan tribesmen: "What must those poor diggers have thought of the funny people from the north country who swapped inestimable salt for stuff whose only role was to give men pride and pleasure by letting them see its luster."[21]

The Edible Rock

Salt was not just pleasing to the taste buds but also an essential electrolyte for making sure people remained energetic and attentive, and as a cure to "heaviness of mind."[22] But perhaps the quality of salt most valued was one indirectly shared with gold—the amazing ability it had to prolong the life of valued commodities. While gold gilding could preserve metals by insulating them from attack by oxygen, salt protected essential edibles from attack by bacteria. In a world without refrigeration, the power to preserve food was a priceless charm. Preserved meats and vegetables could help communities endure famines, make seasonal climatic changes less consequential, and provide high-protein diets for militaries.

The stark irony of salt was its ubiquity in the oceans, where it was least useful to humans, and its relative scarcity on land. Indeed, the presence of salt in the oceans prevented human civilizations from harnessing the most essential resource for survival—water. Sailors languished from thirst while surrounded by water, even as on the coast, elaborate evaporation mechanisms were constructed to gather salt from the sea. In a strange twist of resource inefficiency, water from such processes, despite the need for it, was often lost when it evaporated. Desalination at a larger scale did not come into being until the seventeenth century, when Japanese sailors began to purify water using earthen retorts.[23] Conversely, the coastal evaporation industry was busy collecting salt, largely with solar power, for many centuries before it occurred to anyone to find a way of recapturing the water that was being lost. The Venetians mastered the art of harvesting salt from the sea through simple innovations in their maritime city. They constructed a series of artificial ponds of seawater, connected with sluices to allow for concentration of brine to occur in successive ponds, until salt precipitated out of the densest solution. When brine becomes supersaturated to around 26 percent salt (normal seawater is around 3 percent salt), it begins to precipitate out of solution. The salt industry in Venice perfected the art of reaching that precipitation level and contributed phenomenally to the wealth of the city, which nurtured intellectuals, traders, and such famous explorers as Marco Polo.

Salt was a common connection also between the Europeans and the exotic East that Marco Polo visited. The Chinese had their own infatuation with salt, which they often derived from terrestrial brine wells. Outside the city of Zigong in the province of Sichuan, elaborate systems of bamboo piping were constructed as early as the eleventh century, and even today the city is best known for its salt museum.[24] In his travels Marco Polo noted that Kain-du salt cakes, stamped with an image of the Khan, were used as currency, even more significantly than the paper for which Chinese technology is most often celebrated.

Brine wells, which we now associate with salinity and waterlogging in developing countries, were at one time a prized find. Long before the age of oil, people were drilling for salt with as much ardor as that exhibited for any other treasure. Ironically, oil was found as a byproduct of such drilling, but often little use was found for it, except perhaps in medieval Persia and Azerbaijan. Here, alongside salt, oil was collected in pools, and spontaneous fires from gas-venting mud "volcanoes" were collectively worshiped by locals for centuries. Yet this spiritually charged oil still gave salt little competition as a commodity. Even in North America, brine wells produced salt for decades before commercial uses for oil were discovered. In 1815, oil was being recorded as an unwanted byproduct from brine wells in Pennsylvania, not far from the site of the first celebrated commercial oil strike by Colonel Edwin Drake in 1859.

Salt trading centers worldwide became crossroads of commerce in the most unlikely places. The fabled city of Timbuktu in the heart of the Sahara, in present-day Mali, was built with salt trade from the mines of Taghaza, a city five hundred miles away, which was known for having so much salt that buildings were constructed out of the mineral. Gold was frequently traded in exchange for salt from the mines of Taghaza, and the wealth was put to good use in building universities and schools around Timbuktu.[25] All this salt from the desert mines of the Sahara headed north for centuries, until more widespread salt mining began in central and eastern Europe during the Middle Ages.

The most astonishing salt deposits in Europe were found in the thirteenth century in present-day Poland and are still the largest active salt mines in the world. Visitors to the great salt mine at Wieliczka in southern Poland, near Krakow, are treated to gigantic chambers that rival church halls with artwork crafted out of rock salt: huge chandeliers, a statue of Pope John Paul II, and even a replica of Leonardo da Vinci's *Last Supper* grace the caverns of this phenomenal site. Farther south, the salt deposits continue into Germany and even into Austria.

Before Mozart's birth, the city of Salzburg was most famous for its salt rather than its music. Much of the wealth and splendor that allowed prodigies like Mozart to flourish in this alpine town, far removed from the power center of Vienna, can be traced back to salt. In the seventeenth century the archbishop of Salzburg, Wolf Dietrich, made salt mining a priority for the region and was able to dominate the economy. Salt became a major cause of conflict between Salzburg and Bavaria, until eventually the Bavarians prevailed and controlled the city until 1829, when a treaty between both sides allowed for the accession of the city to Austria. According to the terms of the treaty, Austrians were allowed to mine up to one kilometer beyond their border, and 40 percent of mine workers had to be Bavarians. Furthermore, the Bavarians were allowed to fuel their salt pans with wood harvested from the better endowed forests on the Austrian side.[26]

Although wars have been fought for salt, there was a perceived balance to the mineral in many minds. Chemists realized the inherent neutrality of salt, as the product of synthesis between an acid and an alkali. Sour and bitter tastes combining to make a more palatable form gave salt a special equilibrium. The Chinese recognized the sincerity of salt by linking it to the mystical balance of yin and yang.

Mohandas Gandhi's minimalist lifestyle made room for salt in his struggle against the British. Considering access to salt a human right, he marched to the sea with his minions when the British threatened to tax the commodity. The salt satyagraha is considered a monumental act of civil disobedience in the history of the subcontinent that made its mark on the British Raj. In their declaration of dissent, Gandhi and his colleagues stated: "We believe that it is the inalienable right of the Indian people, as of any other people, to have freedom and to enjoy the fruits of their toil and have the necessities of life, so that they may have full opportunities of growth."[27] The necessity of minerals like salt may be questioned, but the power of invoking the right to their extraction and use has persisted in human societies in even the most self-effacing communities. Gandhi was also careful to point out: "Earth provides enough to satisfy every man's needs, but not every man's greed." However, the connection between greed and need would be challenged in times to come. As we shall see, in many societies it became increasingly difficult to satisfy human needs without resorting to human impulses for greed.

Our Material Imagination

Flint, obsidian, copper, iron, gold, and salt—these are only some of the most pivotal materials coveted for various reasons by ancient societies. Lodestone, by making humans appreciate the importance of the earth's magnetic field, helped explorers navigate the world and realize their most distant desires. But it was not only the materials themselves that inspired civilizations to develop in various ways but also the manner of extraction and processing itself. Such acts of human development are best immortalized in literary traditions across cultures. The mystery and majesty of what lies beneath the earth has captivated literary minds since time immemorial, but some of the writers in the past two hundred years seem to have most acutely grasped the pivotal role played by minerals.

The mines of King Solomon were mentioned as far back as biblical stories, but it took a fictional account by the prolific H. Rider Haggard in 1885 to convey the excitement of their discovery, entrancing an entire generation of readers and beyond. The launch of *King Solomon's Mines* in London was met with tremendous celebration, and reviewers declared it "the most amazing book ever written."[28] Set amid the backdrop of colonial Africa, the story underscored the salience of discovery and the search for mineral wealth in the entire colonial enterprise. How this impulse brought out the best and the worst among the protagonists is a theme that resonated with the average reader on the street in Victorian England just as much as it does today. Books for even younger readers during this time showed the fascination for mineral extraction that the most tender of minds could still absorb. George MacDonald published his most celebrated children's work, *The Princess and the Goblin,* around the same time, and it too was met with resounding applause. The book tells the story of a child miner and his discovery of a fantasy world reminiscent of some of the grand tales of Greek mythology, in which the underworld was most significant. J. R. R. Tolkien considered MacDonald's work an inspiration for his *Lord of the Rings,* a classic work of fantasy that also resonates with earthly tones of mineral extraction. The rings themselves are elemental in origin, and the Mines of Moria provide a dramatic venue for playing out some of the most consequential sequences in the narrative.

The literary connection to our treasure impulse can sometimes be rather obtuse. Consider the perennial children's classic by Frank Baum, *The Wonderful Wizard of Oz,* published in 1900. Baum's father had been an oil entrepreneur, and mineral wealth figured very prominently in his mind and his writing. For instance, the Tin Woodman and his need for oil was a satirical reference to the power of John D. Rockefeller and Standard Oil. Many scholars now believe that Baum conceived the work as a response to a crisis in the U.S. economy at the time, when silver was pitted against gold as the preferred metal of standard coinage. Perhaps the most compelling image in the book is the yellow brick road leading to the phantom Emerald City, which represents Washington.[29]

Fiction's frequent employment of the sinister and savory powers of gems and crystals also carries over into nonfictional traditions. Gems have been prized not only for their physical appearance but also for their magical healing properties. The importance of minerals (from zinc to manganese to iron) in human metabolism is today widely understood and supported by science, but the mystical power of gems was believed by ancient civilizations to emanate merely from their presence.[30] This belief in the healing power of gems can be found across a wide range of civilizations. More than four thousand years ago, the Sumerians of ancient Iraq made reference to healing gems in their writings. Across the Indus River, Vedic scriptures include reference to potions made from crushed gems, which still constitute a major part of Ayurvedic medicine. The Hindu belief in seven bodily energy centers, or chakras, was inextricably tied to minerals and their atomic vibrations. Farther north, the first Chinese medical book, legendarily ascribed to Shen Nong, the fabled Red Emperor, contains detailed descriptions of gemstones and their influence on the body. Although the physical healing power of gems does not have any credible scientific basis, the principles of action may be comparable in efficacy to acupuncture and reflexology.[31]

Many Old World philosophies, including those of the Greeks, the Romans, and the Hindus, relied on five classical elements of nature: earth, air, fire, water, and ideas (referring to abstract energy). The Chinese had a slightly refined version of the elemental sphere that did not include air but did add wood and metals, along with a vital life force, or energy, called *qi*. Feng shui, the ancient Chinese art of geomancy, is based on interactions among these classical elements (fig. 4). Of particular relevance was the significance the Chinese attached to a sodium- and aluminum-rich pyroxene mineral with greenish hues—commonly called jade. Much of Chinese mythology revolves around this "stone of heaven," which was used for the adornment of royalty or crushed into potions that were believed to be elixirs of potency and life itself.[32] Hindu mythology also gives tremendous importance to metals, and there is a continuing tradition of ritual metal purchases on the first day of the annual Hindu festival of Diwali. On this day, which is called Dhanteras (honoring wealth), all Hindus are encouraged by tradition to buy something metallic as a mark of fundamental wealth to honor the goddess Lakshmi, who is believed to control human fortunes.

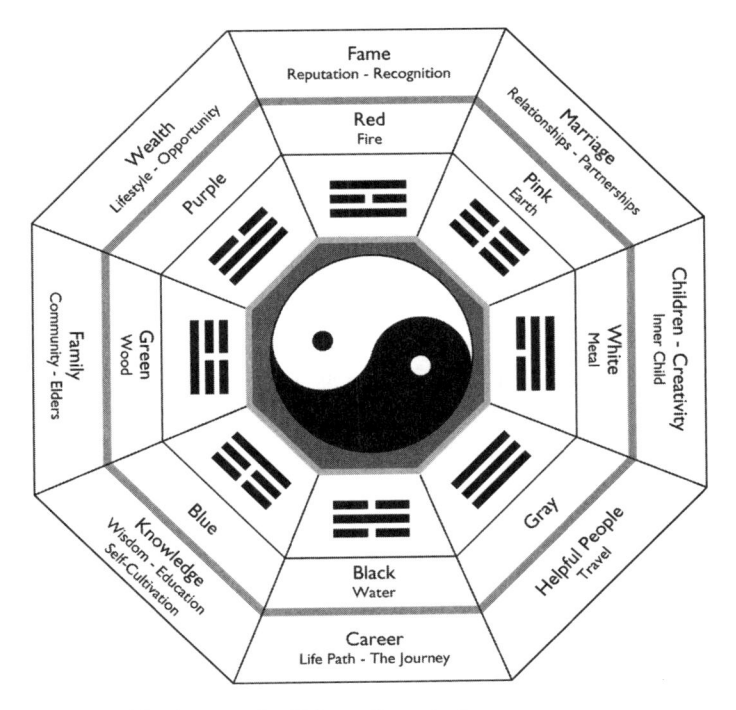

FIGURE 4. Earthly elements in Chinese feng shui

The Chinese have made a culinary connection with rocks that goes beyond the fabled "stone soup."[33] Visitors to the Orient are often intrigued by the colorful displays of mock food outside restaurants. The value of these items and their display goes beyond just advertising or visually explaining the menu. Ambling through the National Museum of Singapore, one discovers that the stones used to construct the fake food models are themselves believed to hold value for the observer. The Taiwanese artist Hsu Chin-I created a "Banquet in Stone" in a remarkable exhibition that highlighted the literal connection between geology and gastronomy. To top off the feast, one is tempted to eat a dark brown jinxiang stone from Yunnan province that naturally smells like chocolate, due to the decomposition and reaction with the stone of certain minute organic materials that accumulate in its environs.

From death rituals to life-giving edibles, minerals are inescapable in our world and have defined both our needs and our wants. Funerals and festivals alike have brought forth their acute importance to us as a species and a society. No matter how much we try, we cannot escape their essential grip on our lives and our futures. Human societies have relied on minerals to develop their infrastructure and cherished them for cultural communion. Minerals have been essential aspects of trade and barter in ancient societies but have also invoked greed and grievance. Some of us have glamorized their discovery, but others have marginalized those who bring them to us from the bowels of the earth.[34]

Our increasingly complex reliance on the elements of the planet is a perplexing attribute of modern times. Through the ages we have valued minerals for their utility and their unique appeal to our senses. Yet we are now at a stage in our history that their appeal as a substance needs to be considered in both ecological and economic terms. From mines to markets, the value that is added to a lifeless piece of rock is a very human story with much vitality for the contemporary consumer. Nowhere is this more lucidly exemplified than through the enduring tale of jewels.

2

CREATING VALUE: THE ENDURANCE OF PRECIOUS JEWELS

> For as long as history has been recorded, precious gems have been searched for and fought over, used to show affection and to create alliances between one government and another.
>
> —Madeleine Albright, speech to the Gemological Institute of America, 2006

Stones are often a metaphor for sterility. We refer to an unfeeling person as stone-hearted or a cold one as stony faced. Yet many of these inanimate natural materials have a concealed beauty that brings joy to people. Scouring through dirt and rock, our ancestors found remarkable secrets that the earth kept, and they learned to prize what they found. Such discoveries continue to this day, delivering great costs and benefits where gems are found. Geologists, archaeologists, construction workers, beach wanderers, and gardeners continue to reveal the earth's rarest treats with various levels of luck. The incidental discovery usually paves the way for more deliberate digging, whether for science or for fortune. The allure of a jewel seems to instinctively leave us entranced and searching for more.

In late May 2000, geologist Javier Garcia-Guinea was digging with resolute purpose in a rather barren part of the Spanish coastline, near the city of Pulpí in the Almería region. There was nothing particularly appealing about the rocks that stood before him: gray, rough, and drab. Yet he knew from his study of the region's crust and some recent calls from local treasure hunters that surprises might be in store for him. Ancient silver and lead mines had existed in this region in Roman times. Soon after World War II, the region's mineral wealth was once again tapped as a result of economic pressures. In 1947, an international economic embargo against the government of General Francisco Franco led Spain to reopen many of its old mines for domestic consumption, and the small mining village of Pilar de Jaravia had emerged out of desolation to enjoy a few decades of prosperity. However, all the ore had long been depleted, and silver was not what Garcia-Guinea was looking for. His prize was likely to be far more dazzling and dramatic, if only he could find the fateful fracture in the rock.

Assisted by a team of adventurers from the Madrid Mineralogical Club, including veteran collectors Angel Romero and Manuel Guerrero, Garcia-Guinea was able to find the promising void in the landscape that had been investigated by local treasure hunters. Some of Angel Romero's colleagues had initially been reluctant to share the potential discovery with a scientist. Their first impulse was to covet their fortune. One of their colleagues, Efren Cuesta, who had been the first to observe the site, would have preferred to create a site on the Internet to sell the treasures that lay beyond the void, but the conservationist impulse prevailed and the scientists were called in.[1]

A dark void of any kind is paradoxically a good sign when looking for geological treasures. The smaller the opening, the more likely it is that something quite special lies beyond it. The team was looking not for a cave but for a geological feature that one usually finds in a museum shop—a geode. Derived from the Greek word for earth, a geode is a spherical hollow rock formation with a crystalline interior. Usually geodes look utterly ordinary from the outside and fit in the palm of your hand. They have to be cracked open to reveal their crystalline splendor.[2]

The Spanish team was trying to find the world's largest geode—one that was made under just the perfect conditions to allow the formation of giant crystals of selenite—a crystalline form of calcium sulfate. Chemically, this compound is quite abundant, and most often the mineral associated with it is gypsum, which is used to manufacture that most abundant of construction materials—cement. The same chemical that forms a relatively unattractive yet useful substance like cement can also take a visually stunning bejeweled form. The right environmental conditions can rearrange the molecules in a way that metamorphoses the same chemical into a perhaps less useful but aesthetically more appealing gem. Consider the case of graphite in your pencil and diamond on your beloved's finger—elementally identical carbon atoms, only rearranged differently by natural processes to give utility in one case and glamour in the other.

In the case of the Spanish geode, the Mediterranean Sea had probably played a pivotal role in this transformation. Dissolved calcium sulfate from the sea, which today is less than two miles away, had inundated the region several million years ago. First, the water dissolved materials from the host rock to create the void in question, and then subsequent seepage of saline water and its painfully gradual evaporation had produced the enormous shards of crystal that embellished the cavity.

As the Spanish team broke open the giant geode, which turned out to be more than twenty-five feet long, and carved an opening large enough for them to enter, they felt the glamour of a movie star. Garcia-Guinea recounted to me in correspondence soon after that he felt like he had entered "Superman's secret cave." Even the man of steel, in that most celebrated of comics, cannot help being attracted to jewels and crystals that constitute his abode at the North Pole. The creators of the Superman saga, Jerry Siegel and Joe Shuster, realized as early as 1932 that crystals would add an indelible allure to their stories.[3]

Pictures of humans inside the giant geode of Almería were released soon after its discovery, and they told a tale as wondrous as the fictional planet of Krypton. Although he couldn't resist the elation of literally sitting in the midst of this natural jewel, Garcia-Guinea followed the scientific impulse to find ways of first studying its form and understanding its origins. In cooperation with Spain's Ecological Police, the environmental authorities of Andalucía and the Council of Superior Scientific Investigations, Garcia-Guinea had the area sealed to public entry. His next step was to look for chemical clues regarding the origins and age of the crystals, which he continues to undertake as proposals are considered to allow the public to enjoy the charms of this natural wonder.[4]

The Stone's Story

Just as the Spanish geode had a story to tell of its origin at the point of human contact, the resplendent jewels that have graced cultures and individuals have their own stories to tell, in many different ways. No doubt the geological history of a stone is its most enduring legacy. Molecules of various salts may have congealed under volcanic heat, then been transported through tectonic forces, or perhaps precipitated or crystallized as the weight of water receded from some deep fracture in the earth. Yet the most consequential story for humans may unfold from the point of contact between people and nature, and what happens to the stone as it is chiseled out of its resting place in the earth. Whether it ends up in a geologist's pocket or a treasure seeker's satchel, a discovered stone will have to contend with human desires. It may become the center of attention for a small mining community and move through multiple hands while lifelessly providing for an exchange of resources between traders. Might it lead to conflict or convey sentiments of joy between lovers that seek to mark their affection? It may accidentally be lost along the way or stolen, leading to anguish for the one who lost an opportunity at fortune, celebrated as a testament to luck or prayer by the one who found it in a random encounter. The stone will surely have a story to tell that goes far beyond its geologic origin.

Surat, India, may seem an unlikely place for the stories of gems to intersect. Hindu nationalism runs strong in this metropolis, and the legend of the war god Indra's jeweled net might well be an apt metaphor for the fortunes of this city with a population of more than 4 million. The story goes that Indra has a special net in which every thread that intersects has a perfectly faceted gem that reflects the facets of each and every other gem in the net, thus creating a network of inordinate complexity.[5] The metaphor appears to take a far more literal twist in Surat, for this is where more than 85 percent of the world's gem-quality diamonds by volume, and 60 percent by value, find their way from mines as far away as South Africa and Siberia.[6] Imagine the story that each of those gems might tell, from where it came and to where it will go once it has been cut and crafted.

Situated two hundred miles north of Mumbai, at the mouth of the Tapi River in the Indian state of Gujarat, Surat had the reputation of being a rather squalid settlement and was the scene of a pneumonic plague epidemic in 1994. The city had a tumultuous past since the earliest days of its establishment in the sixteenth century, when colonial powers jostled to control access to a harbor area that was far enough from the largest cities to not stir local resentment. It was near Surat that the Portuguese were defeated by the fleet of the British East India Company, which managed to secure its most significant concessions from the Mughal emperor Jahangir.[7] The Portuguese then looked westward to satisfy their treasure impulse, and a hundred years later they found the next major global diamond deposits, in Brazil.[8]

Diamonds had been a major enticement to the East India Company, as the only known source of the gems in the seventeenth century was the famed mine of Golconda, a few hundred miles inland from Surat on the banks of the Krishna River. It was in these ancient mines that such fabled diamonds as the Koh-i-Noor (Mountain of Light), Darya-e-Noor (River of Light), and the mystically blue Hope Diamond had been found—their discoveries enshrouded in Hindu mythology and tales of regal splendor. European accounts of this enthralling location date to the writings of the Paris-born gem trader Jean-Baptiste Tavernier, who was commissioned by Louis XIV to find the best and the brightest gems. In his six voyages to India between 1631 and 1669, he documented that as many as sixty thousand diggers could be found in this area after each rainy season, when the diamond-bearing alluvium was deposited on the floodplain. The tradition of digging here dated back thousands of years, as references to diamond deposits could be found even in ancient Hindu texts. The diggers operated under an erroneous assumption about the origin of diamonds, often considering a piece of quartz to be an unripe (*kacha*) diamond that would ultimately grow to become a ripe (*pakka*) gem.[9] Diamonds were the ultimate fruits of the earth for those who embraced this metaphor in the fertile fields of Golconda.

The chemical nature of diamonds eluded geologists and chemists for centuries, and many came to conclusions similar to those of the illiterate miners of Golconda. Eminent gemologists and geologists, such as Max Bauer, as late as the nineteenth century believed that diamonds grew in or from quartz or a form of sandstone called itacolumite. This inference was the result of the common fallacy of assuming causality by common occurrence. Since quartz and itacolumite were fairly common and hence often found around diamond deposits, the geologists simply assumed there was a transformative connection between these materials.

As a result of his experiments with diamonds in the mid-seventeenth century, Sir Isaac Newton suggested that they were carbon-based compounds, similar to wood. This idea was considered impossible until the British chemist Smithson Tennant examined the weight of carbon dioxide emitted from the burning of diamonds versus the burning of charcoal, which confirmed Newton's insight that diamonds were indeed the purest form of carbon. Scientists were once again led astray, however, by even the veracity of this finding. Since diamonds were now known to be carbon-based, the scientific community therefore assumed that their origin was organic—theories were propounded that plants produced diamonds over time, which would make them a renewable resource! They elegantly proposed that perhaps wood decayed over time to first peat, then coal, and ultimately diamonds. Since their form resembled droplets, some researchers in the nineteenth century pondered whether diamonds might be a very pure form of amber, which was known to be the fossilized resin of trees. The paleontologist Johan Goppert studied diamonds with a microscope and saw in his mind's eye fine green needles, leading him to declare that diamonds were the product of algae, so he named two species of algae after the Latin name for diamonds: *Protococcus adamantinus* and *Palmogloeites adamantinus*.[10] The mystique of the most precious diamonds was thus given a strange botanical twist.

Although the stories of the gems of Golconda have been the subjects of entire epic monographs, it is even the smallest gems that come through India's diamond sector that reveal the flow of wealth that holds such consequence for all those along its path. Amid the narrow streets of old Surat one encounters young men brandishing white envelopes that contain brownish-looking ungainly little stones—often tiny diamonds in the rough. Each stone in those envelopes may have traveled thousands of miles to reach this juncture, where its fate will be irrevocably changed by the skill of the cutter and polisher. Each trader that carries one of these envelopes would undoubtedly have some connection to a foreign land, for Indian mines are no longer actively worked. Most likely, the trader would have found the gem somewhere in Africa, since 65 percent of the world's diamonds come from that continent.[11] The largest producer of diamonds is Botswana, which mines them on an industrial scale, making it improbable that any of those stones would be in the little envelopes of the street traders of Surat. Instead, it is more likely that the stones in these envelopes made their way from the informal mining camps of Angola or Congo, where an itinerant miner might have dug, sifting through piles of pebbles in a shallow river, hoping to find the dull sparkle of the stone in the muddy washes of alluvium on his sieve (hence the term alluvial diamonds). He may well have taken what he gleaned to a nearby town and met a trader, quite likely of Lebanese descent, who bought it from him for what might seem a handsome sum to the miner, who could feed his family for perhaps a week or maybe even a month with the income. The Lebanese trader would then have probably encountered someone based in an African capital, possibly a relative of one of the traders in Surat, who buys from the more adventurous rural traders. Once the stone reaches the capital, it soon lands on a flight out of the country, typically via some major commercial metropolis like Dubai. Crossing the Arabian Sea by plane or by boat, it would make its way to India.

On a flight from Kinshasa to Paris in 2004, I was told this typical story of a diamond's travels by one such anonymous dealer—scores of journalists who have covered the trade in diamonds will recount similar map lines for a diamond that reaches Surat.[12]

The stones from the larger mines would still find their way to the cutting and polishing chambers of the city, but most likely through distribution networks of the larger cutting firms that frequent the more upscale diamond-dealing venues of Antwerp, Amsterdam, and Tel Aviv, or through commercial contracts with large mining firms like De Beers and Rio Tinto. Unlike the alluvial diamonds, the stones from the large mines will be more likely to carry a certificate with the title "Kimberley Process," which states that they are not from a location where the proceeds from their sale might be used to finance armed rebellion or human rights abuses. The certificate will have its own story to tell, and one that still may be disputed. The Kimberley Process for diamond certification applies only to rough stones, and there is always a chance that stones may be smuggled from a noncompliant state to a compliant one and get certified there as locally mined gems.

The origin and supply chain for diamond jewelry has become one of the most contested arenas of social activism. There is little doubt that profits from looted alluvial diamonds have fueled civil wars in Angola, Congo, Liberia, and Sierra Leone.[13] Writings on the topic are highly polarized, with such contentious titles as *Glitter and Greed, The Heartless Stone,* and the *History of a Cold-Blooded Love Affair*.[14] Activism about the role of diamonds in sparking conflict came from a very clear linkage established by various United Nations Security Council panels that examined the causes of conflicts in Africa in the late twentieth century.[15] Many hundreds of years ago, the Vijayanagar and Bahamani empires had fought over the diamond fields of Golconda, and now critics argued that diamonds were causing similar resource wars. The fall from grace of this coveted gem became an appealing topic for journalists, and even Hollywood was enticed into producing *Blood Diamond,* a film starring Leonardo DiCaprio.

Douglas Farah, an investigative journalist for the *Washington Post,* has suggested that diamonds from Africa might also be linked to the funding of Islamic terrorism, given the convoluted connections between Lebanese traders and African gem markets. He based most of his accounts on a series of meetings with particular Arab traders linked to the bombing of a U.S. embassy in Kenya while he was a correspondent in West Africa. Although Farah's description of how money might be stored by individual rogue traders, as exemplified by the flight of Taliban capital to Dubai in the form of gold and gems, is compelling, his larger hypothesis of a shadow economy for terrorism is less convincing.[16] The funding of terrorist organizations by this mechanism has been challenged by Islamic finance researchers, such as Ibrahim Warde, who consider the likelihood of gem exchanges to be remote, given the relatively small cost of many terrorist operations and the risk involved in undertaking diamond smuggling. Warde explains that analysts fail to appreciate that only organized military warfare is expensive, and much of the informal militancy that characterizes Al Qaeda is cheap, so financing is hardly consequential. For example, the terrorist bombings that hit the London transport system in 2005, killing fifty-two civilians, cost less than a thousand dollars; the Madrid train bombings that killed almost two hundred people in 2004 cost less than ten thousand dollars; even devastating attacks like those that took place in the United States on September 11, 2001, cost less than half a million dollars in total planning and implementation expenditure.[17] The skepticism of researchers like Warde is supported by the findings of the 9/11 Commission, which stated: "We have seen no persuasive evidence that Al-Qaeda funded itself by trading in African conflict diamonds."[18]

Even if they are exonerated of such charges, the world of diamonds continues to be elusive, and there are many questions about the role diamonds can play in development. How can these most prized treasures help alleviate poverty rather than concentrating wealth in the hands of those who need it least? Might there be opportunities for resolving some of the most intractable conflicts between diamond detractors and those who covet them for wealth and power?

Diamonds and Development

In Botswana, diamonds tell a rather different story from those of war-torn countries like Congo and Angola. A year after the nation gained its independence from Britain in 1966, diamond deposits were discovered in Botswana, and the country's income has increased tenfold since then. Industrial diamond mining operations have transformed this largely dry southern African country from one of the poorest on the continent to the wealthiest; its per capita income of more than fourteen thousand dollars is the highest in Africa.[19] All of Botswana's citizens enjoy free public education, from elementary school all the way to a doctorate, owing largely to the huge diamond royalties, and the literacy rate is over 80 percent. As measured by the United Nations' human development index, a broad-based gauge of the well-being of a country's citizens, Botswana's people are better off than those of almost all other African nations. In general, countries with a formal, large-scale base of organized diamond mines score better on this measure than those with more informal mining sectors, which have clearly fared much worse in development indicators (fig. 5).

The country does have its share of challenges, including one of the worst AIDS epidemics in the world; some 22 percent of the adult population is infected with HIV. Thanks to its high income and educated establishment, the country was the first to provide comprehensive nationwide antiretroviral therapy for patients with AIDS. It is also a leader in biodiversity conservation of rare ecosystems, such as the Okavango Delta.

In spite of the country's wealth, there are concerns that diamonds have produced "distorted development" in Botswana. The economy has not diversified enough to allow for larger job-creation opportunities, and the highly qualified Botswanans who graduate after a free education still have difficulty finding jobs, leading to a 25 percent unemployment rate.[20] The government has tremendous power and is the largest employer, controlling

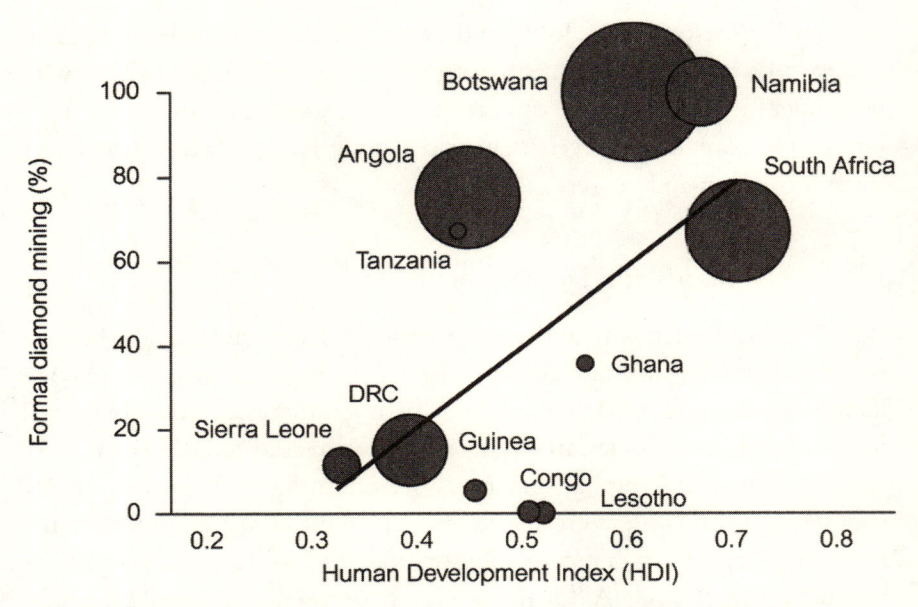

FIGURE 5. Formal diamond mining (as a percentage of all diamond mining) in selected African countries, and individual well-being as measured by the U.N.'s human development index. The size of the circles represents total annual diamond production. (De Beers Report to Stakeholders, 2007, based on data from the U.S. Geological Survey, U.N. Development Program, and Partnership Africa, Canada)

45 percent of the jobs in the country. At the same time, the mine workers know that 70 percent of the country's export revenues depend on their work, and the unions have tremendous leverage on setting national priorities. In the view of John Holm of the University of Botswana, the nation has an "overdeveloped state, an underdeveloped private sector, school graduates leaving the country to find work, a submissive civil society, and a skilled working class of diamond miners who could bring the whole structure down."[21] Yet one is left to wonder by this line of reasoning: would the situation have been better if diamonds had not been found?

In recent memory the predicament of the Kalahari San population in Botswana led many detractors to claim that diamonds had eroded social justice in the country. Environmental and social activists from Europe and North America complained that the indigenous San population, who had led nomadic lives in the Kalahari desert, were being forcibly moved to allow for diamond prospecting. The government had argued that the decision was purely based on finding ways to provide the San with education and health care. The San people had a special resonance in Western popular culture, because of their representation in numerous documentaries as well as the hit feature film by the late South African director Jamie Uys titled *The Gods Must Be Crazy.*

The removal of the San population was litigated in Botswana through local civil society organizations, and on December 13, 2006, a three-judge panel in Botswana's high court ruled that the eviction of the San from their diamond-rich game reserve was illegal. Regardless of the merits of the case, this decision was at least significant in helping to dispel the cynicism many Westerners have expressed about judicial systems in Africa, and also about the power of diamonds in corrupting the governance process. The court defied the common perception of judicial cooptation by the ruling elite. More than a thousand San who were moved to urbanized camps reminiscent of Native American internments are now free to move back to their ancestral lands. The government's case for removal was predicated on grounds of education, health, and acculturation that mirrored many of the justifications given by the U.S. and Canadian governments in the nineteenth century. The court in Botswana has averted the forced acculturation predicament whose legacy continues to haunt many Native Americans.

The desolate Kalahari desert has been at the crossroads of conflict between environmental conservationists, miners, and indigenous rights activists. While the outcome of the case in favor of the San is laudable, the conflation of these various conflicts has generated more heat than light on resource policy in Africa. To environmentalists, the Kalahari San epitomized the image of the "noble savage" being driven to the periphery by modernity and consumerism. An easy target for activists was the diamond giant De Beers, which holds sizable shares in Botswana's lucrative diamond mines along with the government. The checkered history of De Beers in South Africa and its involvement in antitrust litigation in the United States (which was resolved only in 2005), made the accusations compelling. However, there was little connection between the San removal and diamond concessions. Indeed, as Nicholas Oppenheimer, the chairman of De Beers, argued in a rare article for *The Ecologist,* having a local San workforce would have helped the company if diamonds were to be extracted across the game reserve.[22] Linking the San case to diamonds was a useful strategy to gain public sympathy, but in the long run it may lead to further confusion about the role of minerals in development.

Diamonds, while being a luxury item for many in the West, are a source of basic prosperity for the people of Botswana. This is perhaps a rare tale of the wants of the wealthy in developed countries providing for real needs in a developing country. The challenge for mineral economies like Botswana's remains one of wise investment of such windfall capital flows. Developing robust sustainable and diversified economies from diamond revenues will be the real test of the country's leadership. As for the Kalahari San, their decision not to embrace modernity must be respected just as much as Americans allow the Amish or the Mennonites their choice of lifestyles. Perhaps the diamonds of Botswana will overcome the stigma of conflict when they bring not only prosperity but also pluralism to their land. The court ruling showed that this is indeed possible.

Botswana's relative success as a diamond producer has been a subject of considerable study by political scientists.[23] No doubt good leadership played a central role in allowing for the country to prosper after colonialism. Sir Seretse Khama, whose name means "the clay that binds together," was the first president of Botswana, and is often credited with instituting a mature political culture in the country, by encouraging respect for the rule of law through personal example. The country's relatively small population of about 2 million and a land mass the size of France made resource distribution manageable. The dominant ethnic majority, the Tswana (80 percent of the population), have generally been conciliatory toward minority groups, and where conflict has arisen, institutional processes have addressed the matter adequately, as exemplified by the San case.

In Botswana, wealth generated from diamond mines has permeated a mature society, and this has a positive impact on development indicators. The country has used its fortunes to conserve environmentally sensitive areas like the Okavango region, home to the world's largest inland delta system. The influx of revenues from ecotourism and the service sector has the potential to diversify the country's economy in the long term. Consumers who purchase diamonds can feel some sense of comfort that, in at least one African country, they seem to have made a positive difference through acting on their desires.

We might be justified in asking if that same level of positive wealth could find its way to the squalid slums of Gujarat, where our diamond cutters in Surat may well be polishing a diamond mined in Botswana. Many of the diamond dealers of Surat are a rare breed. More than 50 percent of the dealers come from a small religious community called Jains, who exhibit remarkable development indicators. The Jains have a literacy rate of 94 percent, compared with the national average of 65 percent; female literacy exceeds 90 percent (nationally the average is 54 percent).[24] Known for their pacifist political persuasion and their absolute vegetarianism and respect for life, the Jains consider it a sin even to step on an ant. They developed their skills as diamond entrepreneurs because of the community's association with the rulers of the small principality of Palanpur. Because of the Jains' strong religious ethic and dependability, the ruling family of this principality gave them the task of crafting gem jewelry. Once they had acquired these skills, the key was to find a way to best leverage them in a global setting. Many of the eastern European diamond traders of Jewish ancestry felt a common bond with the Jains as a religious minority, and when India opened up to investment, the cutting and polishing sector began to blossom.

The connection between these two disciplined ethnic minority groups is evident in the name of one of the largest diamond companies in Surat, a firm called Rosy Blue.[25] According to the founder, Dilip Mehta, the name of the company was derived from a popular Indian coloration for a blue gem, but when he mentioned this name for registering the company in Antwerp in 1973, the registrar thought he was saying "Rosenblum."[26]

The Jains may have mastered the art of diamond cutting also because their disciplined lifestyle and attention to detail was particularly well suited to the profession. In early Buddhist tradition, which is shared by many Jains, there is a notable text, the *Diamond Sutra,* that is perhaps called by this name because of the allegorical connection to resilience of the gem, as well as the patience and diligence of the diamond cutter. The short text, which can be read in less than forty minutes, was so popular during the heyday of Mahayana Buddhism in the Tang dynasty of China that it was printed. The oldest known dated version of any printed text (considerably older than the Gutenberg Bible) is a copy of the *Diamond Sutra* in the

British Museum, with a confirmed printing date that corresponds to the year A.D. 868. The controversial Tibetan Buddhist practitioner and jewelry entrepreneur Geshe Michael Roach, who wrote a best-selling book titled *The Diamond-Cutter: The Buddha on Managing Your Business and Your Life*, claims great inspiration from this text, which also infuses Jain businesses. The metaphor of the diamond is used in two ways, according to Roach: first, the complete clarity of a perfect diamond symbolizes transparency and emptiness that has the latent potential to be filled with virtue; second, the diamond's resilience symbolizes the ability of ethical practices to endure all. In his opening chapters he tells of his encounters with one of the leading diamond-cutting families in Gujarat and a dealer named Dhiru Singh, whose visits to New York for directors' meetings were also spent as a "spiritual fast, praying in his small hotel room over the garish lights of Times Square, late into the night."[27]

Whatever the roots of their success, the Jains and diamond cutters around the world exemplify a tenacity of purpose to a highly demanding profession. Each individual cutter must be trained to masterfully observe a rough stone and craft from it a gem that meets painstaking criteria of its value, widely referred to in the diamond trade as the four "C"s: cut, clarity, color, and carat, or weight.[28]

The eye is of course the most valuable human tool for this process of value determination, and it is no wonder that many of the cutters are very young, often following the path of tennis players by retiring under the age of forty as their vision becomes less acute. Because of the need for good youthful vision, child labor has been a concern. A labor study in 2001 found that, out of an estimated 2 million diamond workers in the area, 5 percent were children as defined by the International Labor Organization, 96 percent of them twelve to fourteen years old; 98 percent were living with their families in Surat.[29] However, Nicholas Kristof and other writers have argued that after-school employment of children in such professions to supplement their income in otherwise destitute circumstances should be considered a valid development tool.[30] What remains to be seen is whether the wealth that has clearly accumulated in the elite families of Surat will percolate down to the young cutters and their families. Only time will tell if the level of affluence that mining diamonds brought to Botswana's population with relative equity can also be attained in the cutting quarters of India, where over a million people depend on the sector for their livelihoods.

Jeweled Capsules of Time

On February 20, 2008, two treasure hunters reported a remarkable discovery near the small village of Deutschneudorf on Germany's border with the Czech Republic. Using geological surveying instruments, they found what they claimed was a fortune that had been looted during World War II by the Germans from a room in the Catherine Palace near St. Petersburg, Russia, and hidden in this distant location in the dying days of the Nazi regime. The chamber from which this treasure was drawn had been adorned with a gem that emanated from a living organism, rather than a lifeless rock, and had been prized for just as long as gold or diamonds. Amber, the fossilized resin from ancient trees, was the dominant decoration of the fabled room that bore its name.[31] The room had been originally constructed in wall panels made of Baltic amber for the Charlottenburg Palace in Berlin, but the Prussian king Friedrich Wilhelm gave the whole ensemble to the Russian tsar Peter the Great in 1716, and it remained in Russia until the Nazi looting.

Amber is formed when a tree, in response to an intrusion or a wound, releases resins as a defense mechanism against insects and pathogens. These resins slowly become a congealed solid mass called copal, which over many thousands of years becomes fossilized and eventually transformed into the more lasting polymers that we call amber. Because the original tree resin is very sticky, small insects easily become trapped in it, where they become entombed permanently. A piece of transparent amber may contain an insect or other organism that remains perfectly preserved millions of years after it lived.

The beauty of amber was appreciated by the ancient Greeks. The philosopher Thales discovered another charming attribute of the material as early as 600 B.C. Rubbing the fossilized material with a cloth in the dark produced sparks of static charge—the first human-observed manifestation of electricity apart from lightning in the sky. Many centuries later, when the British scientist and physician to Queen Elizabeth I, William Gilbert, coined the English word for this phenomenon, he chose the Greek word for amber, *elektron,* as the root for the term "electricity." The power of amber was felt far beyond the static charges that it produced to the amazement of the ancients. It was believed to have healing properties and was mined at a large scale for centuries to meet both ornamental and instrumental uses. Among the world's largest amber mines was the Yantarny (also called Palmnikcen) amber mine in the tiny Russian enclave of Kaliningrad on the Baltic coast. Amber is often found near coastal areas, because the resinous material can float on water and is deposited along riverways and deltas. During the Soviet era the Yantarny mine produced as much as 650 tons of amber annually. Of this amount only 10 percent was jewel-grade amber, and usually 90 percent was of poor quality and used for organic chemical processing after being distilled in enormous iron retorts. More than 60 percent of the distillate was used for a high-grade varnish known as amber colophony; 15–20 percent was used for amber oil, which is used in pharmaceutical manufacturing and casting. Around 2 percent of the residual amount went into distilled acids used for manufacturing dyes. Given this diverse range of products, the mine was of considerable commercial value during World War II, and it also became the site of a labor camp.[32] Toward the end of the war, as the Nazis were being defeated, thirteen thousand inmates from the Stutthof concentration camp were forced to march to this mine. Several thousand died in the march; the surviving detainees were to be walled up inside a tunnel in the amber mine, but this plan was rejected by the mine's manager. Nazi security forces then brought the prisoners to the beach of Palmnicken during the night and forced them, under rifle fire, to march into the Baltic Sea to drown. Only thirteen inmates survived the tragedy.[33]

Amber has itself entombed life from the past and inspired literature in ways that few other gems have. Whether it is a delicate fern or a forbidding fire ant, the spectrum of creatures found petrified in amber makes it a unique gem that helps us reflect on the larger schemes and cycles of living. The great eighteenth-century essayist Alexander Pope expressed his wonderment at the material in his "Epistle to Dr. Arbuthnot":

> Pretty in amber to observe the forms
> Of hairs, or straws, or dirt, or grubs, or worms!
> The things, we know, are neither rich nor rare,
> But wonder how the devil they got there!

A much earlier text, from the first century, by the Roman poet Martial lays out this fascination with amber in comparable terms.

> An ant beneath a poplar found,
> An amber tear has covered round;
> So she that was in life despised,
> In death preserved, is highly prized!

Modern literature has highlighted the value of amber in very different ways. In Michael Crichton's novel from 1990 and the subsequent feature film *Jurassic Park,* dinosaurs were genetically reassembled using DNA derived from a mosquito's blood meal entombed in amber. Treasure seeking was given a whole new scientific twist with this proposition, and many forensic geneticists began to actually consider the possibility of retrieving ancient DNA from amber-entombed insects. In 1992 a team of researchers from the American Museum of Natural History in New York were indeed able to extract DNA from an ancient termite encased in amber at least 20 million years old.[34] Some DNA has also been recovered from plant fossils in clay sediments in locations including Clarkia, Idaho, but the level of preservation achieved in amber is unique, a result of its desiccation and insulation of the organism from oxidation and other chemical changes. Although it is a stretch to think that the minute amounts of DNA extracted from an amber insect could be used to reconstruct an entire organism, the importance of amber in understanding ancient molecular genetics is phenomenal. Great care is needed to ensure that the extraction experiments are carried out under clean conditions to ensure no contamination from contemporary bacteria. In 1995, one team of researchers took such extraordinary precautions to actually revive an ancient bacterium from a stingless bee in amber from the Dominican Republic.[35] While some researchers still questioned the veracity of these results because of the possibility of contamination, the theoretical potential for revitalizing ancient organisms from amber was once again brought to global attention. Even a remote chance that extinction might possibly not be forever is an important environmental hope for the amber treasure seeker.

Another organic jewel that is produced as a result of an intrusion, in this case of a grain of sand or a parasite, is perhaps the most prized of all organic jewels—the pearl. The lowly oyster reacts to an insult of this kind by covering the intruder with layer upon layer of shell-forming tissue called nacre. It is this slow layering over time that gives pearls their characteristic luster, since light is refracted not only from the surface but from the concentric inner layers of nacre within the jewel.

Pearls have been valued for millennia, and references to the gems can be found in early Chinese texts dating back to the Han dynasty (205 B.C. to A.D. 226), as well as in Roman scriptures. Christopher Columbus was given a list of items by the Spanish royalty to bring back from the voyage that led him to the Americas: pearls were at the top of this list.[36] Religions as far afield as Islam and Mormonism have venerated pearls. In the Quran there are several references to heavenly beings having the metaphorical beauty of pearls; and one of the main scriptures of the Mormon faith is called *The Pearl of Great Price*.[37] Even communist China was fascinated with pearls, to the extent that a metallic image of Mao Zedong was inserted into the mantle of many oysters so that they would produce a "Mao Pearl"![38] The Duke of Saxony in 1612 passed a law by which only royalty could wear pearls, which prohibited the wearing of pearls even by nobility, professors, doctors, or their wives who might otherwise be able to afford it. John Steinbeck immortalized the treasure impulse for pearls in his 1947 novella *The Pearl*, which tells the story of a young Mexican who finds a precious pearl that changes his outlook on life completely as his relations with those he loves are strained by materialism.

As with all gems, the value of the substance is determined by its rarity and the rigor involved in retrieving it. Pearls are no exception; in fact, they were especially coveted because they were not terrestrial and required deep diving to be obtained naturally. Unlike some other organic gem materials, such as ivory or coral, pearls could be extracted without destroying the organism itself, and thus had the potential for a more sustainable harvest. Saltwater oysters are not found in shallow waters and require diving in fairly deep waters to retrieve. Freshwater pearls from specific mussels that grow in lakes may be obtained from more shallow water. Before the advent of underwater equipment, skin diving was the main means of obtaining saltwater pearls. This required very specialized skills that were often culturally passed on from generation to generation. The Indian Ocean basin and the Persian Gulf were the traditional pearling waters of the Old World. Long before oil was discovered, inhabitants of modern-day Bahrain and Qatar engaged in pearling for their livelihood.

On the other end of the Indian Ocean basin, the Malay were well known for pearling, as were the Aborigines of Australia. After European colonization the northwestern coast of Australia became the primary pearling center of the world, and was also the scene of tremendous atrocities toward the aboriginal population, who were forced to undertake pearl dives. There was considerable conflict among the settlers about the treatment of aboriginal pearl divers in this region, and Catholic leaders lobbied the Swan River Colony for regulations to improve the working conditions for them. Matthew Gibney, a priest in Australia who later became the bishop of Perth, recorded some of the incidents of aboriginal lament in the industry around 1878: "When the natives gathered together at nightfall I was surprised to hear the heart-rending wail that went up from the crowd. I found on inquiry, that some of the young men who had been forcibly detained on a pearling boat had been drowned on making an attempt to escape from their captors. They believed they could swim to shore, a distance of about six miles, but only one man reached it."[39] The typical pearling fleet at the time required a schooner of fifteen to eighteen tons and as many as forty or fifty divers, who dove out of smaller dinghies, so space was highly limited. The Aborigines were initially coaxed aboard the ships with enticements of tobacco, blankets, and flour, but once they were on board they were locked into a contract from November to April and were no longer free to leave.

The aboriginal predicament in Australian pearling took a remarkable turn with the advent of deep-sea diving suits at the end of the nineteenth century. British-Australian pearlers, prejudiced as they were, considered the heavy metallic suits, often made of copper, too technical for the Aborigines to use, and they shifted their attention to Southeast Asian divers instead. Japanese divers were well known in this profession and flocked to the town of Broome in Western Australia. Even today the Japanese influence can be felt in Broome, with cemetery stones and markers of their contribution to this sector.

It was the Japanese who also got pearling out of the deep and into the shallow waters of aquaculture. Near the end of the nineteenth century,

Japanese researchers began to experiment with seeding oysters to form "cultured pearls." A young greengrocer from the town of Toba in Shima province of Japan, Kokichi Mikimoto, and a Japanese government biologist named Tokichi Nishikawa were the first to grow cultured pearls on a large scale. The perfection of this process allowed growers to produce mostly spherical pearls, which revolutionized the pearling industry. Thanks to appropriate branding strategies and the exotic appeal of the Orient, Western consumers embraced the cultured pearls with aplomb. Despite the acrimony of World War II, pearls were a source of mutual pride for the Americans and Japanese in Okinawa, where Mikimoto had established one of his oyster farms. Indeed, the Mikimoto pearl was the first Japanese brand to penetrate the postwar market in America.

This breakthrough for pearls also transformed the gemstone sector, by legitimizing the concept of cultured gems. Consumers began to value the shape and form of a jewel, even if human intervention had played a more critical role in its production than in the past.

Values and Tastes

Deconstructing human desires to understand why certain gems may be deemed precious as compared to others is a challenge for even the most discerning marketer. Just as cultured pearls became fashionable, synthetic gemstones that are chemically identical to their naturally occurring counterparts, and perhaps more perfect in their crystalline form, might one day be in equivalent vogue. Yet the sources of natural gemstones continue to be sought by collectors at the highest level, and many laboratories all over the world are busy doing analyses to figure out where a gem might have come from. Using complicated analytical tools, like inductively coupled plasma spectrography, they try to understand the signature spectrums of light passing through various kinds of geological formations and typical impurities in deposits. The task is arduous and expensive, but the collector who is interested in a pigeon-blood red ruby from Burma or a cornflower blue sapphire from Sri Lanka may still be willing to pay the premium.

The analyst's job is to try to find the origin in a way that does not damage the gem, thus defying the claim of an eminent mineralogist that "the best way to identify a mineral is to destroy it, and any mineral worth its salt should welcome being scratched and smashed as the price to pay for being recognized."[40] Even after this rather painstaking effort, the results might still be inconclusive. Gemological purists have often looked at source association of gemstones with contempt, since a beautiful gem in their eyes must be judged on its merits of creation in nature or in a lab rather than geographic location. For many years the Gemological Institute of America, which developed the global diamond grading methodology, refused to give any source for gemstones. The former president of the organization, Richard Liddicoat, believed quite fervently that "if it was a good stone why should it matter where it came from?"[41] Furthermore, for many alluvial gems the geological signature might not be definitive anyway, since the stones could have been transported hundreds of miles from their source by rivers that flow across national borders.

Yet even if the origin is not accurate, the treasure hunter might still want an earthly connection. Thus, cultured pearls might not be analogous to a lab-created ruby because they still come from a natural, albeit caged, oyster while the synthetic ruby is made within the four walls of a lab room. This may explain why the synthetic gem market has not taken off in the way it was expected to. The technology to manufacture synthetic rubies has been available since the French chemist Auguste Verneuil invented the flame-fusion process more than a hundred years ago. Synthetic diamonds have also been manufactured for at least half a century on an industrial scale (not gem quality).[42] A major antitrust case brought by the United States against De Beers was in fact related to charges of price fixing for synthetic industrial diamonds in concert with General Electric, which had patented the process for converting graphite to diamonds under high pressure. As the gem-quality diamonds began to be produced, *Wired* magazine predicted in 2003 that diamonds at five dollars per carat were only a few years away![43]

This has clearly not happened, though a market for synthetic gemstones has arisen in its own right within specific sectors. The branding of "created" rubies, emeralds, and sapphires is now gaining momentum, even as the value of natural stones remains strong. Assembled stones, such as opal doublets and triplets, which provide a "veneer of reality," have remained steady but are no threat to the full natural opal market. Perhaps the most likely area for synthetics to flourish in the near future will be in the realm of colored diamonds. Only one in ten thousand natural diamonds will earn a fancy color grade, and an attractively colored diamond can be several times more expensive than a colorless diamond, selling for as much as $200,000 per carat. In this arena, synthetic diamonds might have the potential for creating a niche. Marketing is pivotally important in this regard. Pink and light brown colored diamonds were initially not highly prized, but when production started at the Argyle diamond mine in Australia, preferences began to change in the mid-1990s. The mine had naturally occurring brownish diamonds, which the marketing team cleverly began to promote as "champagne colored"—and the dynamics quickly changed with this new branding of the stone.

For ecologically conscious consumers, the question to ask might be what the difference in impacts is between mined versus synthetic diamonds. An estimate of energy usage in the production of a synthetic gem versus a mined gem can provide a rough comparison of the relative impacts of both kinds. The problem remains that there is tremendous variation in energy usage among mines, depending on where they are located. The Argyle mine in Western Australia, for example, has a fuel usage of 4.2 pounds per carat, whereas the Diavik mine in northern Canada uses fuel at 11.5 pounds per carat, thanks to its harsh climatic conditions as well as the geological location of ore bodies. Furthermore, Diavik has to produce its own electricity onsite from diesel fuel, because it is so far from the usual sources of power.[44]

The energy usage for synthetic diamonds on a per carat basis may still be considerably greater than for mined diamonds because of issues of scale. Energy use is again highly variable, however, depending on the site. The average energy needed to produce a medium-quality synthetic diamond of three to seven carats at Apollo Diamonds is around 28 kilowatt-hours (kWh) per carat. In contrast, the Argyle diamond mine uses on average 7.5 kilowatt-hours per carat, while Diavik uses 66.3 kilowatt-hours per carat. De Beers, which has a diversified mining portfolio that also includes marine diamond mining off the Skeleton Coast of Namibia, consumes an average of 80.3 kilowatt-hours per carat.[45]

As a mark of our tantalizing treasure impulse in the realm of gems, the cycling of organic materials via synthetic stones has been given an additional twist by a company called Life-Gem. Based in a suburb of Chicago, the company claims to use the ashes from a cremated corpse to create synthetic diamonds. Life-Gem brands itself as a maker of "certified, high-quality diamonds created from the carbon of your loved one as a memorial to their unique life, or as a symbol of your personal and precious bond with another."[46] This patented process, then, gives a new meaning to the marketing slogan "a diamond is forever."

The allure of gemstones has led human beings to venture to remote parts of the world and created markets for substances that might have minimal practical use but immense psychological appeal. The anatomy of this desire has been created by a marvelous confluence of marketing, artistic expression, and cultural connectivity. Their extraction may well have left an indelible mark on the planet in many ways, but there is little doubt that the creation of this market also provided livelihoods where few other alternatives may have existed. The process by which human beings are driven to clusters of treasure worldwide has a history of developing subsequent economies and human settlements that may often occlude their roots in the restless human desire for seeking treasures.

3

THE RUSH FACTOR: TRACING THE MINERAL ROOTS OF GLOBAL POWER THROUGH GOLD, COAL, AND OIL

> It would be a fascinating task to rewrite Economic History . . . to conjecture whether the civilizations of Sumeria and Egypt drew their stimulus from the gold of Arabia and copper of Africa, which being monetary metals, left a trail of profit behind them. . . . [I]n what degree did the greatness of Athens depend on the silver mines of Laurium—not because the monetary metals are more truly wealth than other things, but because by their effect on prices they supply the spur of profit.
> —John Maynard Keynes, *A Treatise on Money*, 1930

Waiting in traffic during rush hour, one may often wonder why people flocked to a particular location to build a metropolis. Why are particular areas the hub of human activity and the intersection of so many professions? The physical location of cities in terms of their environmental and social impact is increasingly important for environmental planners to consider in this age of carbon footprints and water scarcity. Yet often the location of a city was determined by the same treasure-seeking impulse that leads people to explore the remote ends of the planet for minerals. What leads people to develop cities in one place versus another? Might it be the climate, availability of water, or the physical beauty of the landscape? More likely, a city would arise due to economic expediency that has historically been rooted in what is being produced by the land in its proximity. Whether agricultural production was an essential prerequisite for the formation of a city, or whether agriculture evolved as a consequence of urban settlements, remains a contested question by historians and planners.

The European historian Paul Bairoch argued that agriculture was an essential prerequisite for human communities to rush toward a particular location and then consolidate their settlements for purposes of efficiency.[1] There are good reasons for trying to understand the origins of this urge to urbanize: most advancements in human societies have been associated with the development of cities. Even if one assumes that cities were a consequence rather than a cause of human civilizational development, one cannot discount the strong relationship between development of human societies and cities. Once they were formed, cities also became efficient sources of habitation where social services could be easily provided and knowledge between cultures exchanged to further develop ideas. Thus dense human settlement—often characterized by squalor, the spread of diseases, and improper management—was also the site of many technological advances and the establishment of centers of higher learning.

Our ancestors who were hunters and gatherers and searched for more utilitarian treasures at some point started to consider what they could do with surplus resource accumulation. In high-mortality ecosystems or in areas where there was tremendous resource scarcity, such as the deserts of Australia, urbanization was not a preferred choice of habitation. However, a surplus of resources usually spurred some structures of commerce, and cities were the means of moving this forward. Such a surplus of resources was often galvanized by some higher-value mineral wealth.

In *The Economy of Cities,* Jane Jacobs hypothesized that mineral wealth might have been the precursor to the establishment of cities and even agricultural systems.[2] Using the example of a fictitious city called New Obsidian, she recounts how a stock of volcanic glassy minerals used by early humans for tools is controlled and traded with neighboring hunting villages. Communities that lack control of the stock have a demand for the obsidian, so hunters travel large distances to barter their stocks. Some hunters bring live animals and produce to New Obsidian, thereby providing essential food stocks. Obsidian is used as a currency by the residents to acquire materials that are not local. Such rudimentary economic interaction makes the city into a center of commerce, and more people as well as livelihoods are attracted to the center. Seeds of agricultural products might be part of this trade, and some experimental activity is encouraged by the confluence of interactions between various traders. The evolution of agriculture through this process might come about as "the selection becomes deliberate and conscious. The choices made now are purposeful, and they are made among various strains of already cultivated crosses, and their crosses, mutants, and hybrids."[3] This wave of high-yielding cultivation leads to a food surplus, which is partly absorbed by the population increase that results from growth in the labor force that the industries have brought. Agricultural development, in turn, allows New Obsidian to have nonfood imports from other communities. Artisans and entrepreneurs of all kinds capitalize on this profusion of commerce in New Obsidian, and neighboring settlements start to cluster around the city. Such is the anecdotal account of the birth of a Neolithic metropolis, according to Jacobs, whose origins could be traced to a rush for minerals.

Many American cities developed around minerals in unlikely ways—such as through the guidance of domesticated animals, who licked their way to salt deposits along trails. Villages were often built around animal "licks" and sometimes named for the quadrupeds that led them to the precious spot. A salt lick on the road to Lake Erie may be the origin of the name of Buffalo.

The role of excess resource availability in the development of cities was most notably popularized by the Australian philologist and archaeologist Vere Gordon Childe when he coined the terms "neolithic revolution" and "urban revolution," to highlight the role of commerce in the development of cities and their ultimate potential for survival.[4] While some of his work had racial undertones, since it implied that societies that formed cities were somehow more civilized, he forced many archaeologists to ponder difficult questions about human habitations. Subsequent research in many other areas revealed that cities were by no means a European or Asian phenomenon only but in fact were found in all societies where resource availability made it possible to establish them.

Once historians and archaeologists were able to dispense with their prejudices, they found that cities were not as rare as they had previously thought, and indications of such settlement patterns could be found in most human societies at some level. In Africa, for example, ancient cities, such as Meroe in the western Sudanic empires, Ashanti cities in West Africa, the Swahili city-states in East Africa, and Great Zimbabwe in southern Africa, all showed how resources were used to establish complex human settlements.[5] Rural economies around these cities supported the urban center, which also served as a fortification against external attacks. The agrarian periphery was of course essential to sustain the city no matter how it had initially been established.

The Mineral Cities

The rolling hills and remnants of farmlands around San Jose, which are noticeable in a slow rush-hour drive through Silicon Valley, reflect this city's agrarian past as the first metropolis in the Spanish colony of Nueva California established in 1777. Who would have known that a couple of centuries later this farming community would be the technological capital of the world? Yet this giant leap from a crops to a chips economy also had a most significant mineral interlude. Seventy-one years after its establishment, a small town named Coloma, located 175 miles south of San Jose, became the center of attention, and the transformative process took root that would be the most potent catalyst for change. Energy was harnessed in this little town by the flow of water from the hills to operate sawmill operations producing lumber that was needed at the major settlements along the coast.

On January 29, 1848, a mill worker in Coloma named James Marshall was going about his usual duties of keeping a waterwheel running at the confluence of two small rivers when he discovered a small nugget of gold the size of a grain of wheat. The mill owner was a Swiss immigrant named James Sutter, whose fortunes were reversed soon thereafter. The news of the discovery of the metal quickly leaked, and Sutter's property was overrun with prospectors. Such was the dramatic start of the California Gold Rush. The sleepy farming town of San Jose was to become the first capital of the fledgling territory, as it was much closer to where the excitement of exploration was than the former Spanish (and then Mexican) coastal capital of Alta California in Monterrey. San Jose was a convenient resting spot for gold prospectors heading toward the northern or southern goldfields. Real-estate prices in the valley skyrocketed on account of the gold rush, as perhaps the earliest indicator of what was to come in the twenty-first century's rendition of history repeating itself. The French consul to California reported in late 1848 on the situation at Pueblo de San Jose, a few months after the nugget was found, that "a great deal of land has been purchased there and property has increased from a hundred to a thousand percent in value."[6]

It was the sudden rush of humans from across the land that raised the population of the region by 150,000 within two years. This was about the same number as the Native Americans that resided in the region in the middle of the nineteenth century. Only a century earlier their population had been more than twice as large, but disease and colonial oppression took their toll. The change of demographics was particularly striking, because in 1845 California's non-native population had been recorded at only 7,000 individuals.[7] The gold rush brought back the desire of the government in Washington to bring California formally into the American fold. For several years before the gold discovery, President James K. Polk had tried to purchase California as well as Texas from the Mexicans, who had achieved independence from Spain in 1829 and controlled much of the Western frontier. When the purchasing offer failed, the matter escalated into the Mexican-American War. In the Treaty of Guadalupe-Hidalgo that ended the war, Mexico ceded all lands north of the Rio Grande, including California, to the United States. The fateful treaty was signed only one month after the discovery of gold at Sutter's Mill. There was no direct connection between the gold and the war or its resolution, although the confluence of these events was marked by many as a form of providential reward for the victors. As one New England minister is known to have sermonized after the discovery of gold in California and the signing of the treaty: "God kept the coast for a people of the Pilgrim blood. He would not permit any other to be fully developed there. The Spanish came thither a hundred years before our fathers landed at Plymouth; but though he came for treasure, his eyes were holden that he should not find it."[8]

Only a year before the treaty was signed, Americans who came to the erstwhile Mexican territory had considered themselves foreigners; after it, they considered other immigrants who followed the gold trail to be foreigners instead. As historians Robert Hine and John Faragher have noted in their interpretive history of the American West: "Gold rush society was composed of the most polyglot collection of nationalities since Babel."[9] It was perhaps this diversity that played a major role in developing California as a truly cosmopolitan society. Despite the prevalence of prejudice and entrenched xenophobia that many exhibited in the early days of the gold rush, the allure of treasure simply could not be contained, and eventually the xenophobes had to relent. The gold rush swelled California's population and made the case for formal admission to the United States more compelling. The matter of racial prejudice was still a stumbling point, because at the time the Union was split between fifteen states supporting slavery and fifteen opposing the practice. California would be the tiebreaker, and there was considerable internal dissent about the use of slavery for agriculture in the south but surprisingly less interest in using slaves for mining in the north. The mineral rush in the northern part of the state was largely spurred by individual prospecting and small-scale mining rather than the larger mining operations of later decades that require a huge labor force. According to the historian Leonard Richards, the discovery of gold in California produced a mass migration in which settlers who wanted to exclude slavery prevailed, which in turn set the stage for the showdown over the matter in the Civil War a few years later.[10]

California's eventual admission to the Union as a free state in 1850 may also have given rise to the bizarre institution of "filibustering," derived from the Spanish word *filibustero,* which refers to a pirate. The term traces its origins to this period after California joined the Union and pro-slavery politicians encouraged vigilante adventurers, mostly from Southern states, who sought to overthrow governments in Central America as a bypass of proper political procedures within the Union. The term was subsequently appropriated for the tactic of quashing debate through monologues, now enshrined in U.S. Senate Rule 22. Only a three-fifths majority can stop the filibustering in the Senate.

The early Californians were far removed from Washington physically and were quite preoccupied with the establishment of their own legislature to mark the turn in their fortunes after that momentous discovery at Sutter's Mill. The foundations of a new city were to take root on the impetus of the Sutter family near the Sacramento River, where gold had first been found. The city that Sutter established and named after that river of fortune became the state capital in 1855. San Jose returned to its relatively sleepy origin as an agricultural town, until the new argonauts of Silicon Valley arrived two centuries later.

As for Captain Sutter, his life became a proverbial tragedy; despite his success in establishing Sacramento and owning the property where the gold rush began, he gained little economic benefit from the discovery. The gold diggers overran much of his property, and he sought compensation by the federal government for various losses incurred, but eventually his claims were denied and he died in poor obscurity. Commentators on his fate and on this seminal age of gold remember the era with the following words of admonition and awe from Hubert Howe Bancroft, a gold hunter and historian:

> As with some carcass, hidden in sequestered nook, draws every near and distant point myriads of discordant vultures, so drew these little flakes of gold the voracious sons of men. The strongest human appetite was aroused—the sum of appetites—this yellow dirt embodying the means for gratifying love, hate, lust and domination. This little scratch upon the earth to make a backwoods mill-race touched the cerebral nerve that quickened humanity and sent a thrill throughout the system.[11]

Extraction of precious materials may have been a necessary condition for human settlement in many areas, but it was certainly not a sufficient condition to sustain that habitation. The ghost towns that litter the landscape in much of the American Southwest are a stark reminder of what happens when settlements are established without appropriate planning (fig. 6). Although cities like Johannesburg and Melbourne owe their prosperity to minerals, there were some additional factors that propelled them to their status as permanent human settlements. Perhaps it was a mild climate or an enticing harbor that kept people there after the mining boom to establish a plethora of other commercial activities.

Many mineral booms withered into oblivion without leaving much of a trace, such as the Klondike gold rush in Alaska, which seems to have been kept alive with folklore and a few residual tourist attractions for history buffs and itinerant treasure seekers. The small gold-panning operations that run in such areas attract holiday makers who try their luck on the alluvial soils. Many children have their first encounter with our treasure impulse at such visits. A starry-eyed youngster might be seen ecstatically at such venues showing a shiny golden fragment to an elder, only to be disappointed in

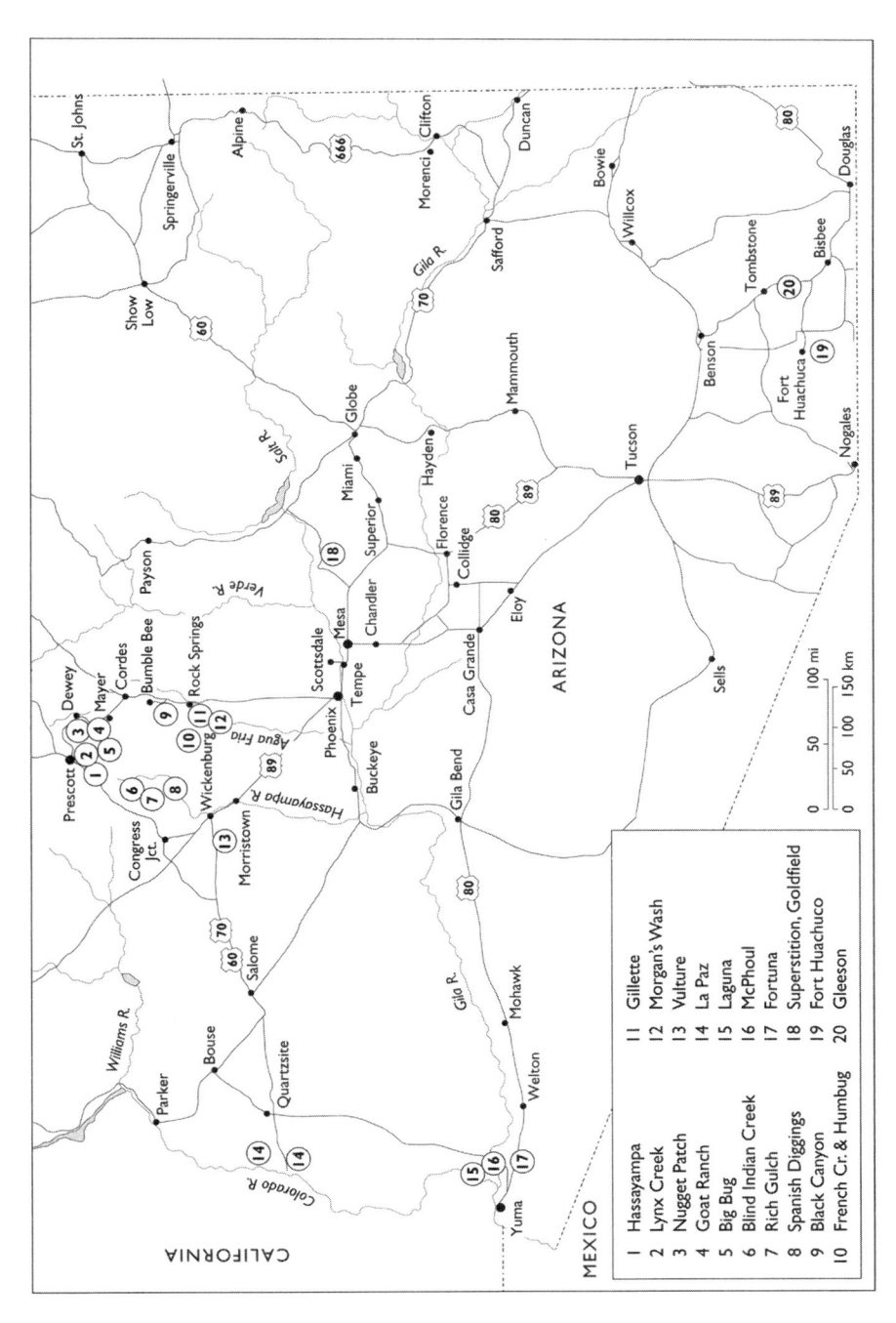

FIGURE 6. A map of abandoned gold-mining settlements in Arizona (Sites identified in Eugene L. Conrotto, *Lost Gold and Silver Mines of the Southwest* [Mineola, N.Y.: Dover, 1991])

learning it is an abundant form of iron sulfide called pyrite (better known as fool's gold).

The quest for gold nevertheless spurred creativity and engineering that were marvels to nineteenth-century commentators. For example, Charles Nordhoff in his 1873 guidebook to California described a three-thousand-foot tunnel near the Yuba River that had been dug at a cost of $250,000 and completed before "a cent's worth of gold could be taken out of its claims." In other cases, dams were built to divert rivers so men could pick gold more easily.[12] While the phenomenon of mineral rushes can generally be considered a classic tragedy of the commons, featuring a rapid depletion of an open-access resource, there was perhaps another side of the story as well. Economic historians Paul David and Gavin Wright contend that in the larger scheme of things the rush toward minerals laid the foundations for lasting development and diversification in ways that are often neglected in contemporary research: "Rapid resource extraction in America was also associated with an ongoing process of learning, investment, technological progress and cost reduction, generating a many-fold expansion rather than depletion of the nation's resource base."[13]

A similar pattern of innovation and industrial diversification spurred by primary industries was also observed in the Scandinavian social democracies that many development researchers now find so impressive. Until the middle of the nineteenth century, Sweden's main export was bar iron. The country's extensive forests were strictly managed to provide for charcoal to heat furnaces that processed the iron. By some estimates this industry's consumption of wood was four to five times larger than the exports of timber until 1854. Research universities and engineering associations, such as the Royal Institute of Technology and the Chalmers Technical School, flourished during the second half of the nineteenth century and laid the foundations for diverse industrialization. This served the country well when iron smelting moved to other parts of Europe with the development of coal- and coke-based furnaces using the Bessemer and Martin processes. Technological innovation continued to accelerate, and the new Thomas process allowed the Swedish iron industry to make steel from the high-phosphorus iron ores of Lapland around 1880.[14] Asea, which later merged to form Asea Brown Boveri, one of the world's largest engineering service conglomerates, traces its origins to around this time. Alfred Nobel's company also flourished during the late 1800s as he explored various materials for commercial use and accumulated a total of 355 patents. His most celebrated discovery, dynamite, had tremendous potential for destruction but also ended up being an essential tool for construction and further mineral exploration.

Innovations were often inspired by resource scarcity. In Australia, miners had to contend with an acute shortage of water, the traditional medium for panning. Inventive miners there developed a method known as dry blowing, in which a pan of "pay dirt" was tossed under a blowing breeze, allowing the finer dust to blow away while the heavier gold settled in the pan.[15] Regardless of the odds against them, the miners found new ways of responding to the treasure impulse. Mining firms were also amazingly adept at forming networks of knowledge across vast geographies and forming global production networks that fostered technological learning at various levels.[16]

Even now, the small-scale mining communities in Africa and South America build mine shafts with decaying wood and improvise sluice boxes for gold extraction from corrugated metal scrap in the middle of mosquito-infested forests. In Asia, the "ninja" miners of Mongolia scour the desolate steppe in subzero temperatures to find the waste rock from a larger mine that might still have some reserves of gold left. Wherever one turns on the globe, the restless search for riches is too hard to resist.

Anchoring Wealth

On the Micronesian Pacific island of Yap (or W'ab as it is known to locals) the most prominent symbols of value are enormous circular stones, called *rai,* with holes carved in them for transport via bamboo poles. For centuries, the value of each donut-shaped calcite stone was determined by its ownership lineage as well as how far the rock had been transported from its quarrying site, which could have been on neighboring islands, particularly Palau, about 250 miles away. The stones were considered legal tender by the community and mandatory in some kinds of exchanges, though now their role is largely ceremonial.[17] What made the stones attractive as a monetary source was the difficulty of counterfeiting such an enormous entity with distinctive quartz characteristics.

Rarity and unique qualities of a material are essential characteristics for a medium of exchange. The huge stones of Yap and a gold ingot are both hard to counterfeit, and the chemical composition of gold has much to do with its economic value. Gold is among the most inert metals and is not easily oxidized; this makes it several times more durable than most other metals. Pieces of gold remain valuable for millennia and continue to entice treasure seekers even to shipwrecks at the bottom of the ocean, because the wealth is resolutely anchored in this lasting material. Because of its relative permanence, malleability, and cultural connectivity with so many different traditions, gold has always been an important medium of exchange.

Money started out as commodities, which individuals exchanged for the other commodities they needed or wanted. This barter system was succeeded by "representative currency," in which notes are issued in lieu of goods that are stored elsewhere. Minerals, specially silver and gold, were an important means of anchoring wealth in this form until the start of the twentieth century. Due to its relatively universal appeal, gold was a mark of trust that backed currency all over the world in the nineteenth century, when the international gold standard was dominant. Yet as trust developed between economies and financial institutions began to have greater faith in the government, "fiat money" became dominant. Gold and minerals served as a trust-building mechanism for global trade and economic stability.

Since 1971, the U.S. government has disengaged from basing the value of its currency on its gold bullion reserves. As a result, there is an enormous stockpile of gold residing in national repositories like Fort Knox or in vaults of international financial institutions, all of which is perceived to be serving no meaningful purpose and could supply world demand for gold through controlled releases. According to estimates by the World Gold Council and the U.S. Geological Survey, in 2007 there were about forty-two thousand tons of gold left belowground, and more than three times that amount aboveground (fig. 7).

The gold in repositories is likely to be in a fairly pure state and could easily be sold on the market. Indeed, this was one of the reasons why gold was chosen as a standard in the first place. However, in an increasingly globalized world where there is enough mutual dependence and economic trust between countries, that currency convertibility has become an

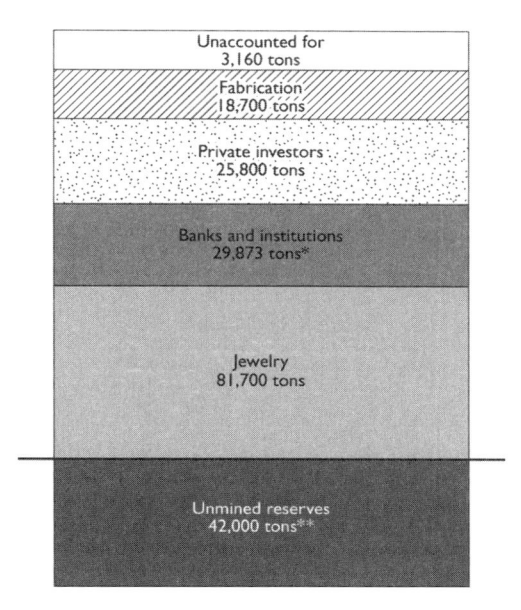

FIGURE 7. World gold reserves above and belowground (Data from the U.S. Geological Survey and World Gold Council; * March 2008 data; ** 2007 data)

Chart contents:
- Unaccounted for: 3,160 tons
- Fabrication: 18,700 tons
- Private investors: 25,800 tons
- Banks and institutions: 29,873 tons*
- Jewelry: 81,700 tons
- Unmined reserves: 42,000 tons**

inexorable trend—indeed the launch of the euro was a hallmark in this trajectory that may ultimately lead to a global currency.[18]

Before the end of the gold standard, many Marxist scholars could blame the wealthy capitalist economies for gold consumption and hoarding, but nowadays the patterns of production and consumption are clearly far higher in developing countries. Industry groups such as the World Gold Council continue to market gold as a sound investment based on their view that, despite the demise of the gold standard, the value of gold has usually not followed cycles of economic decline. According to a report commissioned by the World Gold Council, the exceptional character of gold as an investment vehicle is attributed to three factors: "it is fungible, indestructible and the inventory of above ground stocks is enormous relative to supply flow." Thus, a sudden surge in gold demand can be easily met through gold scrap recovery, having a stabilizing influence on price. Therefore, in times of crisis, the liquidity of gold actually increases. Although the price of gold has fluctuated over the years, and has certainly diminished since its all-time high of more than six hundred dollars per troy ounce, since 1980 it has remained relatively stable between three hundred and four hundred dollars per ounce. This has happened despite the gradual sale of institutional holdings of gold, particularly in Europe.

The durability of gold and its relative ease of recycling pose important ethical questions about the need to mine the mineral, given the potentially severe environmental and social consequences of mining. Should the gold-mining industry transform itself into a material-service provider regardless of whether the material is mined or recycled? Indeed, this is the trend oil companies are beginning to follow by referring to themselves as energy service providers. Whether such slogans are mere rhetoric remains to be seen, but the social obligation to recycle scarce materials certainly falls across the supply chain as well as on consumers.[19] If consumers choose to rush toward gold in times of crisis and hoard the commodity, then recycling is, of course, much more difficult. Unlike savings of cash, savings in gold may pose some latent questions of societal responsibility, since the choice to save in gold indirectly necessitates more mining, which can have serious environmental consequences.

On the other hand, the argument could be made that the choice to have more mining could be potentially positive for emerging economies that have little else in the industrial or service sectors to serve as a catalyst for development. In addition, the livelihoods of the artisanal and small-scale miners who are dependent on revenues from gold should be considered, since there are believed to be several million of them worldwide.[20] However, the relatively small incomes these miners earn, versus the overall profits further up the supply chain, again raise the question of whether a focus on recycling gold and encouraging higher value-added segments of these economies would do more to improve their development path. There is also an uneasy relationship between the artisanal miners and the gold-mining companies, since many of the miners illegally utilize the mineral leases of the companies. Moreover, these small-scale operations are often not endorsed by governments since they get no share in tax revenues. Clandestine mines of this kind are much less likely to have any environmental and safety provisions.[21] While some attempts are being made to mainstream these operations, the systemic question of long-term sustainable livelihoods for these communities remains elusive. These concerns deserve more in-depth empirical study but lead us to further question the symptomatic responses to mining impacts, and rather point toward an integrated economic and engineering approach.

The connection between technological advancement and accelerated gold production also finds its roots in South Africa. Gold is so rare that the portion of ore that represents usable metal is 0.00033 percent, compared with 0.91 percent for copper, 2.5 percent for lead, 19 percent for aluminum, and up to 40 percent for iron.[22] To make extracting gold in any kind of quantity profitable, it was necessary to find a way of separating such minuscule amounts of the metal from its ore. This innovation came in the form of the cyanidation process, which was developed and applied specifically in South African gold mines in the later part of the nineteenth century, leading to a much larger market for mining gold. Gold production in South Africa rose from less than a ton in 1886, when the first discoveries were made, to 14 tons in 1889 and around 120 tons by 1898.[23] According to a former president of the South African Institute of Mining and Metallurgy, "had it not been for the cyanidation process, there is every likelihood that South Africa's economic development would have died before it even had a real chance to begin its true growth."[24] This process, however, led to much more intensive mining, and hence far greater environmental stresses. From an engineering perspective, economically viable and chemically efficient alternatives to cyanidation are arguably close to being developed, but cyanidation continues to be the primary means of industrial gold processing. Artisanal miners use mercury to amalgamate gold from its ore and process it into pure metal, and alternatives to this technique also appear to be elusive. Some technologies have been proposed and developed, but even the United Nations Industrial Development Organization (UNIDO) believes that mercury usage will continue for this purpose in the foreseeable future. The focus of UNIDO's efforts therefore is on reducing miners' exposure to mercury by providing retorts to capture noxious vapors during gold recovery.[25]

Although the cyanidation process has been refined considerably since its initial use, and industry has certain environmental safety protocols for cyanide, there remains tremendous resource usage for large-scale gold mining. Historically, mines required inexpensive labor, which was in ample supply in colonial Africa, where miners to this day work in shafts as deep as twelve thousand feet at temperatures as high as 130 degrees Fahrenheit.

According to one estimate, producing one ounce of gold from industrial mines requires "thirty-eight man hours, 1400 gallons of water, electricity to run a large house for ten days, 282 to 565 cubic feet of air under straining pressure."[26] These enterprises were managed by a ruling elite of European mineral entrepreneurs in South Africa, often referred to as the Randlords.[27]

The relation of gold production in South Africa to consumption of gold in Europe, particularly in Great Britain, is also stark, posing interesting ethical questions about responsibility for contemporary environmental and social issues of gold mining. In his landmark study of the link between the gold producers in South Africa and the Bank of England, Russell Ally shows how the prominence of gold in the global financial system at the turn of the twentieth century arose largely from this connection.[28] The study also reveals that the minting of gold coins and bullion, which could have provided added value to the African economy, was controlled in London, and consumption patterns across the British Empire were managed largely from there. Thus, the demand for gold in this case was spurred in Britain, and responsibility for impacts can be ascribed to British policies. This does not preclude local culpability but shows how we need to be more discerning in who is held accountable in supply-and-demand relationships.

Historically gold consumption was largely intertwined with colonial expansion, but contemporary gold consumption is occurring in many former colonies of their own free will. Consumption trends are often blamed on cultural imperialism, through the advertising and branding of products like fast food or designer clothing, but the same cannot be said of gold. Instead, gold consumption in developing countries reflects the resurgence of latent demand with deep indigenous roots. The three largest gold consumers in the world today are the United States, China, and India—and two of them are the world's largest developing countries (fig. 8). The rapid rise in gold consumption in India during the 1990s is particularly striking and can be attributed to the deregulation of gold pricing.[29] Gold ornamentation has strong cultural roots, but the quantity of gold consumed and the propensity for aggrandizement of the metal is now particularly acute in the developing world.

The Indian case deserves further attention because of the significance of gold in matrimonial customs, particularly dowry. Gold jewelry is regarded

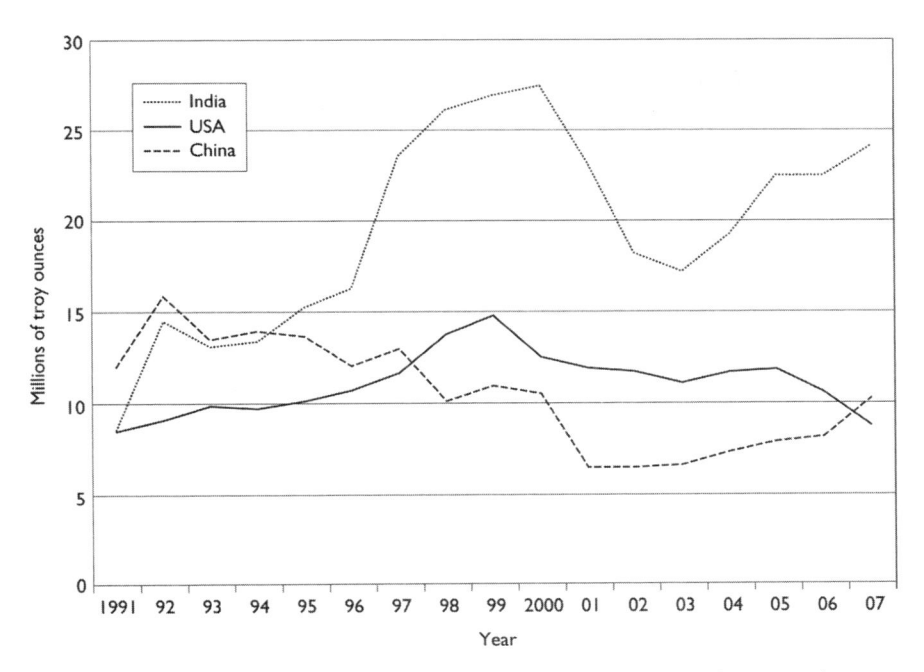

FIGURE 8. Gold consumption trends among the top three consuming countries (Data from World Gold Council)

as a family heirloom and is passed on from generation to generation. For women, gold is an important personal asset, which is gained not only through the recycling of old jewelry but also through additional acquisition. Gold has traditionally been an important constituent of dowry, while in other cases gold has been given to women as a gift that accounts for an asset reserve in patriarchal societies. Yet often the asset is controlled by the men in the household, so even a positive intention for this gift may not be realized. Often gold given to women as dowry is a sort of fee to win a man's hand in marriage. Several activist groups in India trying to fight traditional matrimonial customs are allying with anti-mining activists.

Historiographic work on dowry reveals an interesting colonial connection as well. Veena Oldenburg in her book *Dowry Murder: The Imperial Origins of a Cultural Crime,* has argued that the commodification of dowry, particularly gold, occurred during the British Raj, when land tenure became the unit of capital accumulation for Indians instead of the traditional practice of land use and access rights.[30] Since property was allocated to male heads of household and the use of land became "subservient" to ownership, women were essentially disempowered. Previously, women would always retain land use as a right wherever they went after marriage, but the delineation of property precluded that for them when they moved to their husband's domain. Their sole worth was thus relegated to dowry, particularly gold. However, this wealth was largely static and seldom exchanged for merchandise, as it was meant to be passed on to the next generation. Occasionally the gold was used as assurance for credit, but for the most part the wealth that accumulated was largely barren in terms of its impact on human development.

In the words of Robert Triffin, a Belgian economist who was a major critic of the Bretton Woods system of maintaining gold reserves to back the U.S. dollar: "Nobody could ever have conceived of a more absurd waste of human resources than to dig gold in the distant corners of earth for the sole purpose of transporting it and reburying it in other deep holes, especially excavated to receive and heavily guarded to protect it."[31]

The gold standard may have served some purpose in times of immense global distrust. However, its utility in an increasingly globalized world has waned, even though individual acquisition of gold as an asset remains justifiable. In contrast to the relatively sterile consumption of gold, there was another subterranean secret that attracted great rushes of human activity and was also put to immediate use in energizing the world.

The Omnipotence of Fossil Fuels: Coal, Oil, and Gas

Carbon has been the seminal element of life. The most complex of all compounds are made of carbon and called organic—giving rise to an entire branch of chemistry that bears this name. These organic compounds in all their complexity can ultimately create life forms, or organisms. These organisms metabolize energy from the sun, and perhaps from other sources in the case of deep-water organisms, and upon their death sequester that back into the earth. Seeking out these sequestered sources of energy has been an age-old human preoccupation.

The quest for fuels first drove treasure hunters to forests in search of wood for charcoal and the right stones to ignite that most prized ancient manifestation of energy—fire.[32] Next came coal, which the ancient Chinese in modern-day Shanxi province are believed to have extracted as much as ten thousand years ago for ornamental purposes.[33] The shiny black mass that they were using for embellishments had the same chemical lineage as charcoal and formed over millions of years from the decomposition of plants, not too different from those that made wood fuel. Through the process of fossilization, the extraneous materials of the wood were purged and a high concentration of energy-rich matter preserved, first in the form of peat and then, with pressure and dewatering, eventually turned into coal. Yet a few millennia would pass before coal began to be used as a fuel.

Romans encountered coal during their invasion of the British Isles, where earlier Bronze Age people had used the material for cremation pyres. Initially, the Romans used the shiny black material for jewelry, calling it gagate (the etymological predecessor if the word "jet," as in jet black). They also used it as a fuel to keep them warm and for the furnaces of blacksmiths. Nevertheless, coal's full potential as a fuel was not realized until the industrial revolution, when large amounts of transportable energy were needed to power the steam engine. The proximity of coal mines to some of the major manufacturing centers in the British Isles may explain why industrialization and subsequent global economic dominance took off in Europe rather than China, despite the latter's enormous reserves of coal, which ended up fueling its growth two centuries later.[34]

Although coal continues to be a major energy source, its dominance since the height of the industrial revolution has declined considerably, being replaced by another product of decomposition under slightly different geological conditions—petroleum. Whereas most terrestrial plants lead to the formation of coal, marine zooplankton and algae are believed to decompose into oil and natural gas. The vast expanses of oil and gas reserves in the Persian Gulf are a sign of the submarine past of that region, and indeed some of the more promising discoveries of petroleum in the future are likely to occur under the oceans. The rush toward oil and gas is thus likely to move offshore as supplies dwindle on land.

Flammable oils were known as a very useful source of movable fuel before the discovery of large amounts of petroleum in the United States. Ironically, the first commercial oils that transformed our energy usage also came from the oceans—not from minute lifeless decaying zooplankton in the form of petroleum, but from the heavy blubber of the world's largest living creatures, whales. Whaling had been widespread, from the Vikings to the Aborigines, for millennia, but the functional use of wax and oil from the cranial cavity of sperm whales or the blubber of most species was discovered only in the seventeenth century. Whale oil for lighting was expensive—a gallon of oil in the early 1800s cost the equivalent of two hundred dollars—so it was largely used by the elite or in high-quality public illumination sites like lighthouses. Demand grew in the nineteenth century to such an extent that an otherwise renewable resource was in danger of being irreparably depleted. One of the first international environmental agreements made following World War II was the International Convention for the Regulation of Whaling in 1946.[35]

The discovery of petroleum as fuel came at an opportune time for the whales, and perhaps the international whaling convention got more political support as well because blubber was becoming more dispensable. The fear of depletion that galvanized action on the whaling convention has only recently been raised in the context of petroleum. For much of the twentieth century, humans have used petroleum blithely, and any supply constraints have largely been framed in political terms, such as the oil shock of 1973. Yet the inherent nonrenewability of petroleum and its associated fuels such as natural gas is now beginning to raise concerns. Peak oil—the point at which production from finite reserves will inevitably begin to decline—has become a frequent refrain in the context of energy policy and planning. Although financial analysts have always advised us to diversify our portfolios, somehow this logic eluded the global energy portfolio, which has been excessively reliant on fossil fuels.

Clearly, the dominance of coal, oil, and gas is a relatively recent phenomenon in world energy consumption, but it appears that their ostensible abundance has led us to a sense of plenty that is quite unprecedented in history. The fear of depletion has somehow been eclipsed by a treasure impulse that more reserves will be found in the depths of the ocean or the frigid bitumen plains of Athabasca. The problem with this view is the physical fact that fossil fuels are simply not renewable on human timescales. It is true that technology may allow new sources to be discovered, but ultimately we will reach a limit with fossil fuels, and alternatives will need to be found. However, this inherent obsolescence of oil might not necessarily mean that we should abandon the quest for fossil fuel resources or completely eschew their consumption, since the resource is still providing a real service in terms of energy availability and potentially acting as a catalyst for development as well. We should certainly plan effectively for their depletion but not dismiss their contributions to providing human civilizations with unprecedented access to energy. At the same time, the impacts of resource extraction in the short and long term on ecosystems (both directly through pollution and indirectly through global change processes, including global warming) must be factored into our planning process.

It is essential that we consider resource substitution in an integrative framework that treats human development and ecosystem processes in all their complexity. Danish economist Sven Wunder found counterintuitive examples of oil-rich countries, such as Gabon, that had minimal deforestation, while others like Ecuador had a more checkered record of forest loss. Because of its oil income, Gabon had a reduced incentive to engage in commercial forestry, so forests and biodiversity were conserved, thereby also providing some measure of carbon sequestration.[36]

Understanding the dynamics of such complex systems is at the core of effective planning for resource depletion, yet discourse on the matter has been polarized by a sense of either panic or complacence. The minds of some observers are concentrated on peak oil leading to a "long emergency"; others are avid Cornucopians trying to convince us that technology can somehow provide a "bottomless well."[37] The public is bemused but perhaps also a bit anxious about how to make sense of such conflicting arguments. The 250 percent rise in oil prices between 2000 and mid-2008 certainly led to anxiety and a reduction of SUV sales, but the overall consumption patterns have not changed much. This has led economists to propose that the demand for oil, given our extensive reliance on the fuel, is relatively inelastic. To add to the confusion, prices suddenly dropped back by 50 percent within three months at the end of 2008, but SUV sales remained low because of the overall economic downturn.

Elastic demand provides relief to a policymaker in the way that an elastic waistband provides comfort to an athlete. When demand is inelastic for an inherently scarce resource, policymakers face a quandary of whether to ration supplies and extend the finite spectrum of availability of the goods, or to allow the depletion to run its course and perhaps wait for higher prices to eventually encourage conservation while investment is made in alternatives and technologies. To plan for such strategies, we need to have some forecasting mechanism that is effective in understanding the rise and fall of a resource, but our treasure impulse can often skew the planning horizon in various ways.

While most natural phenomena that involve resource usage tend to follow a pattern of rise and fall, with one distinct peak, extractive resources can still spring some surprises, leading to multiple peaks. New discoveries of oil or the advancement of technology that allows previously unextractable reserves to be harnessed can prolong the availability and make the fall more gradual than the rise. Yet the ultimate decline of a nonrenewable resource is inevitable, and eventually we will need to either develop alternative fuel sources or change our lifestyles to accommodate different patterns of consumption.

Beyond Depletion

Off the western coast of Greenland was the world's first and last cryolite mine, which closed in 1987. It is unlikely that any other source will be found, as the mineral is extremely rare. It also has unusual properties that make it a curiosity for collectors. In water it becomes nearly invisible to the human eye, because its refractive index is almost identical to that of water, so it looks like magically sunk ice in a beaker. Yet the properties that led humans to mine cryolite were far more consequential. Chemically, cryolite is a compound of sodium, aluminum, and fluorine and does not itself have any extractable element to speak of. However, this marvelous mineral played a pivotal role in harnessing aluminum from its most intractable and abundant ore, bauxite. Because of its own unique molecular configuration of one aluminum atom nestled between three sodium and six fluorine atoms, cryolite acted as a flux that helped break the stubborn bonds of aluminum and oxygen in bauxite. It was cryolite that demoted aluminum from a precious metal used in jewelry and the adornment cap of the Washington Monument, in the nineteenth century, to a flimsy wrapping material for sandwiches.

The abundant uses of aluminum in this day and age, from airplanes to soda cans to computer cases, would suggest that the depletion of cryolite signals an imminent warning or "a long emergency." Yet few people outside the frigid frontiers of Greenland even know that large reserves of natural cryolite have vanished. Why was this mineral extinction so uncelebrated in the annals of resource depletion? The simple answer is synthetic substitution. Humans were able to find a way to chemically synthesize cryolite with relative ease, so the demise of the mineral remains unsung. Might the same be possible for some of the other minerals and their derived materials that we so desperately need to sustain modern society?

Technological optimists hope that the same synthesis potential will lead us to salvation in the face of oil, gas, coal, and other fuel depletion. They frequently remind us that the first law of thermodynamics states that energy can be neither created nor destroyed (except perhaps in the nuclear processes in star cores where mass is converted to energy via Einstein's elemental equation). Therefore, the concept of running out of energy is contrary to the laws of physics. Two such Cornucopian researchers, Peter Huber and Mark Mills, assert this kind of refrain in their aptly titled volume *The Bottomless Well: The Twilight of Fuel, the Virtue of Waste, and Why We Will Never Run Out of Energy*. They call environmentalists who are concerned about conservation "lethargists," arguing that such views are a cop-out and that technological progress is the ultimate panacea. The authors claim that efficiency is futile, since any attempts to achieve it will lead human impulses to consume even further, given the pricing dynamics of supply and demand in a free market. They disparage the view that population pressures will reduce livelihood potential and resource availability, pointing to the growth of employment in spite of increases in productivity.[38]

All of this sounds quite glib and might be dismissed by many environmentalists as yet another conservative diatribe against conservation. But there are many influential individuals who continue to believe in this primrose forecast. The front cover of Mills and Huber's book has an endorsement from Bill Gates, the founder of Microsoft Corporation and someone who embodies a very service-oriented treasure impulse. The software industry invented livelihoods where mechanization in manufacturing took them out, and Gates thus has much reason for optimism. However, might the arguments from the Cornucopians be making the same cognitive leap that some resource determinists such as Jared Diamond have made about selective vignettes of history in specific cycles of civilizational collapse? If all has been well in the past, with human ingenuity saving our species in the aggregate, what assurance do we have that it will be the same in the future? Does it not behoove us to be cautious and plan for the worst, given the essential aspects of nonrenewability? In trying to answer these questions about mineral usage and depletion, it is first essential to dispense with the aphorism that "history repeats itself."

In the larger scheme of their studies, physicists will generally agree that the universe is moving toward a greater state of chaos, and we humans are trying desperately to claim some order across a time continuum. Our rates of depletion of resources and the speed with which we replenish depleted stocks with innovations will determine the shapes of any particular resource peaks across this continuum. Furthermore, along these curves we will need to contend with underlying human urges to procreate, possess, and celebrate. How governance systems manage these urges and balance individual choice versus civilizational planning will determine the trajectory of these curves. The acquisitive and creative energy that a mineral rush creates eclipses concerns for our long-term progeny, often termed intergenerational equity. Might we just muddle through the future with spurts of resource rushes, bursts of creativity, and cycles of proximate and contained collapses, followed by ultimate progress?

In the summer of 2007, the Pew Charitable Trusts encouraged and supported visits by journalists to central Canada's vast boreal wilderness. The trusts had been established by the same family that owned Sun Oil Corporation many years ago. Perhaps as an act of atonement, the aim of the Pew visit was to allow writers to observe the world's most ambitious industrial project, which the trust's Climate Change Program considered one of the most significant and preventable contributors to global warming this century.[39] Climate change had already become a household word to describe the impact of a fossil-fuel based economy on the world's climate, because of an imbalance in the amounts of carbon dioxide and other heat-absorbing gases being released versus their sequestration by plants.

The journalists were to witness the initial stages of a mineral bonanza that was targeting fifty-four thousand square miles of a hitherto little known kind of fuel source—the Athabascan oil sands. These extraction projects aim to stretch the depletion curve a few more decades or perhaps a century by exploiting a far more problematic source of oil than Saudi sweet crude—bitumen. For all practical purposes bitumen is tar—the same stuff we use to pave roads. In the heart of Los Angeles, one finds the same oozing tar immortalizing the remains of prehistoric mammals in the form of the La Brea tar pits. The tar is perhaps the only visible reminder of the oil boom that fueled the growth of Los Angeles in the early part of the twentieth century, when oil wells lined Long Beach.

Although not endowed with the same climate as Los Angeles, the town of Fort McMurray in Alberta also originated from an oil boom. A tiny rail stop became a city with divided highways within a few years to cater to this newest global rush for oil sands, and it now boasts a population of more than seventy thousand. James Howard Pew had been one of the first to consider the wealth of this region in the 1960s as a viable resource, and a new science museum in the city honors his legacy by naming an industrial garden of extraction equipment relics after him.

The scale of the projects operated by Syncrude Canada, the world's largest producer of synthetic crude oil, is phenomenal. The tailings dam around one of the largest oil sands projects stands at three hundred feet, and if measured by the volume of material used to build it is bigger than the largest conventional hydroelectric dam in the world—Three Gorges Dam in China. Yet this dam is essentially a waste management system rather than a source of potential energy itself. By 2010, more than $70 billion will have been invested in developing the oil sands projects in this region and as a consequence in building more of these enormous impoundments.[40]

The promise of wealth from a heavy sandy deposit was largely motivated by a group of scientists, led by Karl A. Clark, from the University of Alberta in the 1920s. Through painstaking fieldwork they made the case to commercial developers that there was a technically feasible way to extract fuel from these deposits. Squeezing out energy from the desolate sands of Athabasca tested the resolve of even the most ardent oil enthusiasts. Off and on the projects languished due to commercial concerns about their viability rather than for environmental reasons. Yet ultimately, the treasure impulse for this controversial prize prevailed, even in current times of heightened environmental concern.[41]

The next treasure seekers might head even farther north to seek the methyl hydrate deposits off the Arctic coast. All such efforts are important because they are buying us time to discover a more lasting solution to our energy challenges. They are also providing livelihoods where none existed. Yet the pace of development remains a major imponderable. As with the gold rush, the economic imperative remains to dig as fast as you can to harness the resource before the price drops below a point where mining becomes unprofitable. In the case of oil sands, a decline in oil prices below seventy-five dollars a barrel would make them uneconomical to mine. The financial incentive is thus to mine fast and not necessarily pace the extraction according to ecological constraints, or for that matter with planning horizons in mind that would allow for the concurrent development of some ultimate renewable fuel from algae or cellulose or a yet-to-be-determined miraculous source of material power.

PART II

TOIL AND TREASURE

I shall borrow from Epicurus: "The acquisition of riches has been for many men, not an end, but a change, of troubles." I do not wonder. For the fault is not in the wealth, but in the mind itself. That which had made poverty a burden to us, has made riches also a burden. Just as it matters little whether you lay a sick man on a wooden or on a golden bed, for whithersoever he be moved he will carry his malady with him; so one need not care whether the diseased mind is bestowed upon riches or upon poverty. His malady goes with the man.

—Seneca ad Lucilium, *Epistulae Morales XVII*, circa A.D. 30

4

THE DARKER SIDE OF FORTUNE: THE PSYCHOLOGY OF TREASURE DEPENDENCE

> That consumption is no longer restricted to the necessities but, on the contrary, mainly concentrates on the superfluities of life harbors the grave danger that eventually no object of the world will be safe from consumption and annihilation through consumption.
> —Hannah Arendt, *The Human Condition*, 1958

Traveling through the teardrop-shaped isle of Sri Lanka, one can sense the presence of tea on almost every turn of the road. Terraced hillsides give way to mildly sloping fields of *Camellia sinensis*—the enticing evergreen bush whose leaves have infused calming pleasures to countless generations. The complex array of aromatic compounds within these infusions clearly had some lasting impact on human well-being, to warrant the spread of humble tea from its origins in Asian villages to its obligatory place in English kitchens. Despite their many differences, tea drinkers the world over are bound by biochemical stimuli that are triggered soon after they take a sip.

Of the seven hundred or so chemicals that have been isolated from the tea plant, those called polyphenols make their way to our cerebrum and cause certain neurons to release dopamine molecules into the prefrontal cortex of our brain. This marvelous organic compound signals other parts of our brain and registers pleasure, triggers ambition, and makes us want more of the same. Dopamine is the critical chemical of human creativity, and also of human cravings.[1] Its influence on receptor nerves that register pleasure needs to be moderated by other chemicals in the brain, such as gamma-aminobutyric acid (known as GABA) when it reaches a synapse: that highly consequential gap between neurons where much of the impulsive action of brain circuitry is manifest.[2] Tea, by most research accounts, stimulates only a healthy amount of dopamine and doesn't inhibit too much GABA, so it is more likely to be beneficial than harmful.[3] But we should remain attentive to the delicate balance that the brain must keep for regulating dopamine, because this equilibrium can have a profound impact on the environment and on us in many obscure ways.

Consider the tea fields in Sri Lanka in greater detail. If you were to wander deep into the croplands in the southern part of the island, you would encounter some occasional disturbances in the landscape. Large pits several yards deep will appear in the middle of a field, some of which will be propped up with timber on the sides and often filled with stagnant pools of water and leaf debris. Most of these pits were dug by the same farmers who were harvesting tea the year before. One may wonder why a tea farmer would abandon a stable income from tea cultivation, sustained by human dopamine cravings, and dig into the fields so ravenously? The answer lies in that other fabled symbol of Sri Lanka—the cornflower blue sapphire.

While plowing their fields, tea farmers occasionally encounter a rough gem that they are able to sell for multiples more than a year's tea harvest. Finding this quick wealth short-circuits the dopamine paths in their own neurons and makes them seek even more gems and engage in high-risk behavior that may undermine their earlier livelihoods. When one farmer finds a prized gem, the word quickly gets around and the area is usually overrun with miners from across the region. The landowners may capitalize on the frenzy as well by charging the itinerant miners to camp on their fields. The discovery of a large gem can create such euphoria among the treasure seekers that any rational calculations of probability are largely dismissed. The environmental impact of this restless search for wealth is relegated to a minor consideration for the drooling diggers. Even if the chance is minuscule, the rush to riches prevails among all miners great and small.

The phenomenon is by no means unique to a destitute population—the same impulse is found in gambling behavior and in stock market frenzies that can lead even satiated millionaires to bankruptcy courts. The impulse to take risks spurs ingenuity, ambition, and exploration, but it is also what can lead us to a spiral of addictive and compulsive behavior.[4] Such is the darker side of fortune.

Viral Propensities

Many fortunes have been made at Microsoft Corporation, but of particular interest in our quest to understand the psychology of treasure seeking is the career trajectory of one former employee, whom the company hired in 1981. Richard Brodie was the firm's seventy-seventh worker, and along with the former Xerox executive and software pioneer Charles Simonyi made up the applications division of the company at the time. Brodie was given the task of producing a word processing package for businesses, and he is credited with creating the most widely used software in history—Microsoft Word. The rapid spread of Microsoft Word as a software package, coupled with a personal interest in poker playing, inspired Brodie to research the spread of ideas and human impulses. Meanwhile, Brodie's boss Simonyi amassed millions over the next few years and endowed a professorship at Oxford University committed to a "public understanding of science." The first recipient of this endowed chair was the evolutionary biologist Richard Dawkins, whose book *The Selfish Gene* had been a runaway best-seller in the 1970s and originated a concept that was most intriguing to Brodie.

Dawkins was interested in how certain attributes of cultural behavior get replicated so quickly in societies, and he encapsulated this idea by introducing a heuristic unit of analysis that he called a meme. How do fashion and branding so rampantly take over human populations? Why do so many Harry Potter fans endure queues until past midnight to buy the early copies of a new book or to watch a movie sequel? Why are we so inclined to cluster around celebrities and a narrow range of fashion choices? Humans, Dawkins hypothesized, evolve genetically through genes and culturally through memes.[5] Although he did not go into the biophysical attributes of memes, the concept was intriguing enough to spawn an entire field, "memetics," that was embraced by some as a bridge between psychology and evolutionary biology; at the same time, it was dismissed by others as a pseudoscientific concept.[6] Regardless of this debate, what was clear in the observational world was that human beings have a tendency for taking risks and emulating others in a way not easily explained by conventional evolutionary biology. The matter was not merely pathological, but considering the ubiquity of its prevalence in material culture, fashion, gambling, mineral rushes, and so many other human behavior patterns, it clearly had a "viral" quality.

Business professionals were quite familiar with the powerful concept of "viral marketing," and now it appeared there was an analytical way to consider such patterns of behavior more broadly. Richard Brodie was convinced of the value of the meme concept from his experiences at Microsoft so much that he set out to write a book about it for a general audience. It was completed in 1996 and aptly titled *Virus of the Mind: The New Science of Memes*. Brodie was convinced that understanding our impulses that had strong neurological roots would help us "better navigate through a world of increasingly subtle manipulation."[7] Memes were thus a way to help us understand our impulses and the propagation of desires. In this view, memes were the "software" in our brains that sustained humans through the survivalist period of the species' development, when most were hunters and gatherers. Times have clearly changed, but many of us are still stuck with the same cultural software that has infected us through a process of traditional connectivity with the past.

Consider gambling tendencies more generally. It made perfect sense for ancient societies to engage in high-risk, high-reward behaviors because of a lack of opportunity for gathering survival resources. The impulse was to repeatedly take risks in order to get a lucky break, because the alternative of playing it safe may well have been at least as risky. There was an evolutionary component to this trait, which could be manifest in genes, since those who took risks were able to procreate. However, there was also a more transcendent cultural component that may have spread in our neural circuitry through learned behavior. This component was manifest as "memetic transmission," which had a particularly viral property in that many individuals were "infected" with these traits simultaneously, resulting in their survival. The memetic traits were also more fluid in their transference than genetic traits. Hence risk-taking behavior for food acquisition could pass through our neural circuitry and emerge as risk taking for gambling, or for stock purchases.

Of particular interest to environmental scientists would be the potential for the same survivalist meme in transmitting impulses for accumulating resources. Just as a squirrel stocks up acorns for the winter, humans were well served to acquire as much as they could when encountering a pleasing resource. However, the impulse to store and accumulate food and other precious resources may have been "memetically" transferred to conspicuous consumption behavior, which we have observed in most human societies. The environmentalist John De Graff caught on to the viral metaphor as well when he and his colleagues produced a book and a TV series titled *Affluenza*.[8]

Concepts analogous to viral memes in the context of consumerism are found in many non-Western societies as well when they grapple with needs versus wants in the construction of desire. Consider, for example, the milieu of myths and conceptualizations that emerged in nineteenth-century Mombasa, Kenya, a melting pot for cultures across the region because of its pivotal role as a trading port for the Indian Ocean. The Swahili language developed in this context and reflected some of the ways in which cultural transmission (through memes or otherwise) occurred. Three words described the interplay of human desires in Swahili at the time: *moyo, nia,* and *roho*. Mombasans believed that the roho created cravings in the human mind and was mobile from one person to another—it could thus "possess" individuals through human influence. The moyo acted as a consolidator of thoughts and agency and represented the physical and figurative heart of the individual, which longed not only for objects but also for amorous bonding with other human beings. Although the roho gave birth to wanton desires, the nia was the conscience that attempted to contain those desires and balance the impulses injected by the roho. Mombasans were very conscious of the roho's ability spread its influence, particularly in the context of the ivory trade. In the view of one keen observer of this cultural history: "Mombasans believed when the roho was not restrained, the end was dissatisfaction and social alienation: lafuka or gluttony."[9]

Yet even if various human societies realized the psychosocial perils of excessive consumption, whether aware of memes or not, we are still left with the question of how such a process of unbridled consumer demand spreads so quickly. Consumption on a massive and homogeneous scale needed an agent of change, and this was provided by a fruitful convergence of interests between the information technology industry and the advertising sector.

Richard Brodie was particularly attracted to memetics because it provided a way to understand how mechanisms like technological advertising can take advantage of cultural evolution very effectively. Getting messages across to millions of individuals simultaneously through media advertising has the positive impact of reducing communication time for essential messages to be aired. However, this rapid communication also allowed for the spread of potentially harmful or redundant information about particular products or services without the benefit of processing time and reflection on their value.

At this point it is essential for us to appreciate that consumption has some important biophysical constraints, which environmental scientists have appreciated for decades but are also gaining credence with social psychologists. One of the leading lights of positive psychology, Mihaly Csikszentmihalyi, a co-author of *The Meaning of Things,* states that "consuming consists of energy expended to improve the quality of life by means of increasing entropy."[10] Such an insight is profoundly important in terms of the viral impact of consumption. This ongoing impact inevitably increases thermodynamic disorder, or entropy, in the world, which would require energy of some sort to reverse. Individual actions of consumers through the power of advertising or memetic transfer can have a huge impact in the greater scheme of things. The profound insight of chaos theory has been that even a small change in one part of the system can have an enormous impact worldwide. One of the founders of chaos theory, Edward Lorenz, framed it in the title of his pioneering paper on the topic: "Does the Flap of a Butterfly's Wings in Brazil Set Off a Tornado in Texas?" Memetic transfer follows the same chaotic potential, which was stated most simply by James Gleick as: "Simple systems give rise to complex behavior. Complex systems give rise to simple behavior."[11] This observation may seem fairly self-evident, but it ran contrary to a lot of scientific practice. In ecology, for example, there has been a longstanding belief that nature finds "balance" and equilibrium if left on its own. However, the insights from chaos theory suggested that nature is itself far more complex rather than linear, and disturbances, whether through natural disasters or erratic consumption patterns, can have profound and unpredictable consequences. In the words of ecologist William Schaeffer, who had a conversion from the old orderly school of ecology to the new chaotic school of systems science, nature is in fact "both exhilarating and a bit threatening."[12] In a more poetic moment Schaeffer noted that "what passes for fundamental concepts in ecology is as mist before the fury of the storm—in this case, a full, nonlinear storm."[13]

Two additional factors that must be considered in our understanding of the psychological aspects of material acquisition are time and power. Through technological change, we have accelerated the potential for simple actions to have complex outcomes, which may also affect the absorptive capacity of natural systems. In addition, we are often engaged in consumption behavior to utilize time in various ways—either for productive work or for recreational activities. The amount of time that people have available for consumption has increased with technological change in developed countries. In one notable study, between 1871 and 1981, hours of work per year in the United Kingdom for full-time male wage-earners fell by 36 percent, and years spent in the labor force fell by 17 percent (leading to a combined decline in lifetime working hours of 47 percent). For each one percent increase in labor productivity per man-hour, lifetime working hours fell by 0.33 percent.[14] In poor countries, the situation is more mixed, depending on the income level. Swedish economist Staffan Burenstam Linder made this case presciently in his monograph titled *The Harried Leisure Class*: "The time devoted in enjoying different consumption goods is as essential to the consumption process as the goods themselves."[15]

Given the acceleration of material availability and the huge selection of opportunities for consumption, much of the time people may be consuming without much forethought as well.[16] In one study, American teenagers being paged at random moments during the day reported that 30 percent of the time what they were doing at that moment was not what they wanted to do, but they could not think of anything else they would rather be doing instead.[17]

Whether consumption is existential or experiential, material usage may still occur and have the same cascading impacts on other human beings as well as on the environment. The rather discomforting aspect of the findings from research in this arena is that our consumption decisions may well be made with less direction, purpose, or forethought than would be expected. In such circumstances, the role of institutions of authority becomes all the more salient. How might government or other controlling forces respond to these traits in human societies? Much can be gleaned from considering world politics in the twentieth century.

Operation Abundance

President Woodrow Wilson acknowledged the power of consumption as a means of hegemony most explicitly in one of his landmark speeches, in 1916 to the World Salesmanship Congress. Much of Europe was embroiled in World War I, and Wilson was quite concerned about conflict resolution and global governance, but with an American edge. He stated to the congress that "America's democracy of business" would prevail in "the struggle for the peaceful conquest of the world." His business focus at the time was on palpable consumer products that could "force the tastes of the manufacturing country on the country in which the markets were being sought." As a scholar as well as a policymaker, Wilson also pontificated on the larger ramifications of using the psychology of "tastes" more constructively by stating that "the great barrier in this world is not the barrier of principles but the barrier of taste." By homogenizing tastes, Wilson implied, we could somehow rid the world of at least one prime contributor to our conflict with foreigners and "convert them to the principles of America." This same line of reasoning prevails even today, in Thomas Friedman's Golden Arches Theory of Conflict Prevention, where he somewhat simplistically proposed that countries with McDonald's restaurants are less likely to go to war with each other due to a homogenization of cultural proclivities through fast food.[18]

The evangelism of consumption for a greater good resonated strongly with our species, mainly because it already had a treasure impulse. Consumption in this way came to be perceived as a societal positive and made us all far more vested in and dependent on the social system that it sustained. I was alerted to Wilson's speech by Victoria de Grazia's notable book *Irresistible Empire,* in which she comments that President Wilson was "the first leader to recognize that statecraft could find leverage in the physical needs, psychic discomforts, and situations of social unease being unleashed by the new material civilization of mass consumption."[19]

Three decades after Wilson, and after yet another world war, we were confronted with a further push toward activist consumption. This time it was to show how modes and strategies of consuming materials could ideologically differentiate the left and right wings of the political spectrum. Having fought collectively against a common enemy, the Soviet Union and the United States soon discovered that they had less in common than they thought, and the world consequently endured five decades of the Cold War. Frigid as the war might have seemed to the military realists, the metal smelters and factory furnaces were as hot as they had ever been in human history. Not only were countries producing weapons at an unprecedented pace, they were also manufacturing consumer goods under the pretext that sheer production of palpable products (often at the expense of services) would give them an edge over competing economies. The Soviets were consuming materials for governmental trophy projects while the United States was consuming them to promote contrasting lifestyles of affluence. Both sides were getting hooked on their consumption of materials, but the American arsenal of abundant consumerism was to prevail globally. Our ability to creatively market our products was clearly a competitive edge.

Appealing to the senses and to human acquisitive desires is America's forte. Capital markets are well suited to nurture such creativity, and with the help of mass media and advertising we were able to entrance the world with our products. The Nobel-laureate economist Gary Becker was among the first to consider that addictive behavior was possible in the context of consumption even in the absence of a pathological condition that might inhibit regulatory hormones. In an essay written with economist Kevin Murphy titled "A Theory of Rational Addiction," Becker built on his earlier work on behavioral economics to suggest that even a gradual implementation of a consumption plan with full information about addiction potential "rationally" leads people to addictive behavior, because of the way people discount their utility (reduce the value) for a product over time.[20] Although much of the research in this regard focused on drug addiction, consumerism and material exchanges also are captured by this provocative insight. By one estimate 5 percent of the American public is clinically addicted to shopping, and around 0.7 percent is addicted to gambling.[21] However, if we were to consider the larger issue of conspicuous consumption and its impact on fashion trends, and a domino effect of influence, the addictive definition becomes much broader and encompasses a far greater demographic. Even between enemies, the common human impulses for acquisition and the propensity for addiction in various contexts remains salient. Additionally, behavioral economists are discovering that human beings often do not learn from their mistakes and tend to repeat certain kinds of irrational behavior patterns because of some visceral impulses that we may not fully understand.[22] For example, what started off as a move toward conscious consumerism about water pollution has transformed into an obsession with drinking bottled water. Sociologist Andrew Szasz refers to this phenomenon as "inverted quarantine," where individually cautious behavior is becoming a societal scourge instead of a means of following the fabled "precautionary principle" of ecological protection.[23]

The sheer scale of current mass consumption is hard to fathom without some visual images, captured with phenomenal artistic flair by photographic artist Chris Jordan. His works include an entire replica of Georges Seurat's famous painting, *Sunday Afternoon on the Island of La Grande Jatte*, made with a chromatic assortment of 106,000 aluminum cans—the amount used in the United States every thirty seconds. Another portrait includes 32,000 cyclically arranged Barbie dolls in the form of a shadowy image of a woman's breasts—the number of dolls signifying the average number of aesthetic silicone implant surgeries performed each month in the United States in 2006.[24]

Sociologist David Riesman observed the Cold War of materialism as early as 1951, when he wrote a satirical essay titled "The Nylon Wars." The narrative concerns a U.S. Army effort named Operation Abundance, in which 200,000 pairs of nylon socks, 4 million packs of cigarettes, 35,000 Toni permanent-wave kits, 20,000 Yo-Yos, 10,000 wristwatches, and an assemblage of other excess consumer items are rained down from the air on Soviet population centers.[25] The goal of this onslaught is to make the deprived Soviet population crave the material goods and rebel against their communist masters. The plot thickens when the Soviets respond with their own air raids of caviar, fur coats, and collections of Stalin's speeches. The Americans are able to claim victory ultimately because of a timely drop of two-way radios that empower the Soviet people to order consumer goods as if from a store catalogue. With an onslaught of electronic products, the electricity grid in the Soviet Union collapses, and the capitalist model prevails. Yet the breakdown of the electric grid remains a loose end in the story, which perhaps unwittingly reflected the darker side of mass consumption with an environmental twist. The rise of polymeric materials such as nylon also occurred as a result of fossil fuel consumption being taken for granted during much of that period in history.

Guns, Guano, and Butter

The Cold War was a time of great introspection for human civilization. Scholars such as Riesman considered the paradox of progress, which left them perplexed. How might we balance goals of development, through sustenance of human livelihoods, with the security imperative that ideological differences had thrust upon the world? Economists, sketching the possibilities of production frontiers for college students in introductory courses, continue to formulate this trade-off quite simply as "guns versus butter."[26] The mineral undertones in this formulation are perhaps missed by many economists, but guns and butter are inexorably linked to a common mineral need that led to many colonial conflagrations in the past. Regrettably, terrorists have also been able to make the connection between the common chemistry of feeding humanity and blowing it up.

Recall the vigilance the Federal Bureau of Investigation had to exercise concerning fertilizer purchases after it was discovered that the Oklahoma City bomber Timothy McVeigh had carried out his heinous act with nothing more than ammonium nitrate fertilizer as the explosive ingredient. The same nitrates that are used to grow corn to feed industrial cattle that make butter were also the limiting factor in making guns and explosives functional. Such is the irony of nitrogen chemistry in certain economic diagrams.

Until 1910, the mass production of guns and explosives as well as fertilizers for agricultural crops was made possible through the mining of nitrates in distant lands. For much of the nineteenth century the world's nitrate deposits were heavily concentrated along the western coast of South America — either in the form of guano deposits (bird and bat feces) or through high desert caliche deposits of sodium nitrate in the Atacama region of Chile.

Amazingly enough, here was a mineral that was literally renewable through biological processes on relatively human timescales. Defecating birds and bats were mineral generators that over time could build deposits as thick as 150 feet. Rocky islands were prime nitrate and phosphate territory, because generations of seabirds, plump with fish meals, would generously dump their nutrient-rich excrement on these outposts. In dry conditions, such as those along the western coast of South America, these deposits would harden and appear ready to be mined.

In this bastion of Spanish influence, the British had created an industrial niche for themselves by controlling most of the world's supply of nitrate mineral through their corporate relations, particularly with the government of Chile. Landlocked Bolivia had a territorial dispute with Chile over some access to this vital resource, and, as the value of nitrates raised the stakes for both sides, a protracted armed conflict erupted in the late nineteenth century (1879–1883) between the two countries. The War of the Pacific, or the Saltpeter War, involved major naval battles and was a defining moment in South American history. Peru also became involved by assisting Bolivia, and initially the United States attempted to mediate this nitrate war in 1880, but without much success. After four years of bitter conflict, a truce was finally achieved in which Chile gained most of its demands and control of the valuable mineral areas in exchange for building a railroad from the Bolivian capital La Paz to the Chilean port of Arica. Under the terms of the truce that was finalized in 1904, Bolivia would have complete access to this port.[27]

Soon thereafter, advances in chemical engineering made mined nitrate less consequential, because an innovative process to synthesize the prized material was developed in Germany. Nitrogen constitutes nearly 80 percent of our lower atmosphere, but finding a way of binding it to oxygen to constructively form nitrates eluded chemists for centuries. The first step in the synthesis would be to form the foul-smelling gas ammonia (which contains a nitrogen atom surrounded by three hydrogen atoms). Two German chemists, Fritz Haber and Carl Bosch, were able to develop a process for synthesis, starting from natural gas (methane) and using various metallic catalysts and operating under high pressure conditions to form ammonia. Capitalist enterprise motivated this effort as well, since Bosch worked for the German company BASF and patented the process with their support. The significance of this discovery was as profound as the ability to smelt aluminum from bauxite, because it facilitated the industrialization of fertilizers at an unprecedented rate. There was, of course, the darker side of this dependence on synthetic fertility, which made us complacent about other ecological limits—not to mention the synchronous rise in weapons ammunition, which also became much easier to manufacture. The synthesis of ammonia allowed humanity's burgeoning population to become even more ravenous eaters and trigger-happy warriors. It would be fair to say that the chemical synthesis of ammonia was more consequential for humanity than the invention of the airplane, the television, or nuclear energy. Without food productivity reaching such unprecedented levels, worldwide population growth from 1.6 billion individuals in 1900 to over 6 billion today would have been unthinkable. The movement away from mined nitrates to manufactured ammonia was a pivotal point that tipped us over into the age of plenty.[28]

Haber and Bosch were duly awarded the Nobel Prize in chemistry for this singular achievement in harnessing nitrogen for productive (and destructive) usage. Our drive for more intensive production of our most fundamental human needs of food and security was thus given a major boost. Any psychological constraints that societies had about growth and expansion on account of food or security limitations were greatly diminished. Technological optimism spread infectiously throughout the world as scientists and engineers tried to prove that they could construct molecules, manufacture food, and build material structures at a magnificent scale. They were successful to a large degree in reaching their goals, and much of humanity was in awe of these accomplishments. They appeared to be proving that population determinists were wrong. However, some rumblings were noticeable about the impact of this material frenzy on society from scholars in other disciplines, who began to gain some attention from the public at large in the middle of the twentieth century.

In parallel to the social critics, environmental conservationists had been raising concerns about human impact on the environment and massive development through the construction of dams or the use of pesticides and industrialization. Yet there was little confluence between the sociological critics of consumerism and the environmentalists at that time. Indeed, some of the earlier naturalists of the nineteenth century were often reclusive pantheists who did not particularly engage with communities or consider social networks in the same light as sociological critics of materialism. Human connections were, however, always of more interest to the general public than a linear concern about nature.

By the time he published "The Nylon Wars," Riesman was already a celebrity intellectual as a result of his earlier best-seller, *The Lonely Crowd*, written with Nathan Glazer and Reuel Denney. The book had provided a sorry commentary on American material culture in a nonfictional polemic that was to land Riesman a place on the cover of *Time* magazine. Much to everyone's amazement, this relatively dense academic volume became a runaway hit, selling more than 1.5 million copies, making it the most successful academic treatise in history. Despite the contemporary critique of the book as simplistic and "pseudoscientific," many of its insights can be found in the work of later quantitative researchers, including Robert Putnam, whose book *Bowling Alone* has been comparably successful.[29]

The underlying theme that appeals to the public in these books is that humanity is being absorbed so ravenously in the acquisitive impulse of material wants that we may be losing less palpable societal goods. This is perhaps the darker side of fortune that all treasure seekers must be willing to concede and plan for effectively.

Affluence and Happiness

If there has ever been a universal aspiration for humanity, it is to attain "happiness." Although historically emotions of disappointment and sadness have also been praised as motivators of creativity, there is little empirical evidence in terms of productive quality to support the poet Shelley's assertion that our "sweetest songs are those that tell of saddest thought."[30] Many a disappointed miner would have claimed fame on that account, but the only prospect there might be is that despondency and crisis provide opportunities for some individuals to consider alternative paths or trajectories, which in turn may lead to fulfillment. Thus, there may be some empirical support for the preceding stanza in Shelley's poem: "our sincerest laughter with some pain is fraught." In general, the pursuit of happiness that was enshrined even by the American founding fathers alongside "life and liberty" has been time-tested in various ways despite the natural peaks and troughs of the human experience.

The quest for this emotional state of bliss has been a subliminal motivator for much human enterprise and its consequent impact on the environment. There is clearly a self-centered aspect of attaining happiness—each individual wants to be contented for his or her own sake. However, the aggregate of individual happiness also affects the functioning of society at large, and we must appreciate this collective good as we consider human impulses for material acquisition. The Nobel-laureate philosopher and literary doyen of the early twentieth century Bertrand Russell grappled with this question in his seminal treatise *The Conquest of Happiness,* where he tried to understand the need for happiness and how to achieve it. An important premise that Russell stated in his correspondence regarding this work was the intrinsic worth of happiness at the societal level: "The good life, as I conceive it, is a happy life. I do not mean that if you are good, you will be happy; I mean if you are happy you will be good." This unidirectional causality that Russell proposed is supported by empirical data, though at the time of his writing Russell was not as concerned with neurological studies or labor satisfaction surveys. Rather he was trying to consider postindustrial concerns regarding technology, materialism, and spirituality. Literary scholars in the Victorian era had grappled with these issues quite directly.

In his acclaimed dialogue essay, "The Critic as Artist," Oscar Wilde stated that humanity likes to "rage against materialism, as they call it, forgetting that there has been no material improvement that has not spiritualized the world." This thread has been picked up by many revisionist scholars of consumerism and material culture who ask us to reflect on the social ties created by gift giving, the livelihoods created by the products produced as a result of conspicuous consumption. In his provocatively titled book *Lead Us into Temptation,* James Twitchell tries to make the case primarily through observational analysis that consumer fashions and branding lead to bonding in an age of individualism.[31] Renegade environmentalists such as Jesse Lemisch have been irritated by many green activists like Ralph Nader whom they feel have "turned their backs on people's reasonable and deeply human longings for abundance, joy, cornucopia, variety and mobility, substituting instead a puritanical asceticism that romanticized hardship, scarcity, localism and underdevelopment."[32] British sociologist Daniel Miller believes that we can "feel sympathetic to the dreadful plight of cosmopolitans who feel they have too many pairs of shoes . . . and they bought their child a present instead of spending quality time with them." Miller considers it "not acceptable that the study of consumption, and any potential moral stance to it, be reduced to an expression of such peoples' guilts and anxieties."[33]

To respond to these criticisms, let us return to Russell's very simple assertion that "good" conduct, which could suggest behavioral attributes like being industrious, studious, morally proper, might not necessarily make you happy. Such attributes might lead to wealth if our system worked well. You work hard and you play hard—the mantra of many Wall Street treasure seekers. The wealth from working hard in turn may allow for consumer spending, which the aforementioned revisionists would contend makes us happier. It is thus worth considering how far this line of reasoning is valid, especially since Russell urged us to also consider the reverse—if we are happy, we are more likely to do well for society. The facts regarding this latter assertion are fairly well established. If people are happy individually (but not to the point of selfish complacence, which can be a temperamental trait in some subjects), they are more likely to perform better in work and be more productive family members and active philanthropists in their societal contributions. There is also a circularity to this premise, as the research shows that philanthropy itself makes people happier.[34] The wealthy hoarder has been shown to be less "happy" in contemporary research than the benevolent one.

Although Adam Smith is better known for his treatise *The Wealth of Nations*, which posits a rather self-centered acquisitive view of human impulses, he was also the author of another notable book called *The Theory of Moral Sentiments*, in which he wrote: "How selfish so ever man may be supposed, there are evidently some principles in his nature, which interest him in the fortunes of others, and render their happiness necessary to him, though he derives nothing from it, except the pleasure of seeing it."[35] This view has been picked up by some economists who work with psychologists, ethologists, and neuroscientists to understand the chemical triggers of human behavior, particularly the cooperative behavior within societies and the possible role of material gifts as a mark of reciprocity to reinforce such cooperation.[36]

As far as issues of wealth, acquisitive impulses, and happiness are concerned, the data are far more complex and deserve some careful consideration as we ponder material desires. "Feeling good" has many technical connotations. Researchers tend to use such terms as "quality of life," "well-being," "life satisfaction," and simply "happiness" to define that viscerally positive emotion that has physical connotations related to nonmental health as well as psychological aspects related to a state of mind and consciousness (which may ultimately have some physical connections as well). The research in this area is voluminous, with popular books by psychologists, economists, and medical doctors claiming to provide resolution to that elusive quest for perennial smiles.

Before considering the ways in which material acquisition, wealth, and other factors can lead to joy, we have to be clear about what control individuals may have over their abilities in this regard. According to social psychologist Sonja Lyubomirsky, the acclaimed author of *The How of Happiness,* each person has a happiness "set point," determined by temperamental factors (which are genetically controlled), that accounts for about 50 percent of reported happiness levels. An additional 10 percent is determined by social circumstances, and around 40 percent is under our control through behavioral choices. Consumption behavior would fall within this 40 percent of attributes that could potentially give us a higher sense of well-being.[37]

So the question remains, where we do have control, what makes us happy? Could wealth and material consumption make us happier? Psychologists have struggled to understand human needs versus wants and their impact on motivation and a potentially consequent sense of well-being. Much of this line of study can be traced to the hierarchy of needs developed in the 1940s by the American psychologist Abraham Maslow. This hierarchy of needs now graces most introductory psychology courses and is often depicted as a pyramid, with physiological needs at the base, followed by safety, love, esteem, and self-actualization (at the apex). Maslow suggested that the higher needs become focused only when the lower needs have been met and the individual has overcome the baser necessities. In this view, a certain quantum of material well-being is essential before other aspects can be realized. While the delineation provided by Maslow is useful, there is now general consensus among psychologists that material well-being is inextricably linked to other factors in the hierarchy and that in comparison across populations, self-actualization may be higher among individuals whose physiological needs are comparably lower.

Clayton Alderfer subsequently developed a nonlinear version of Maslow's thesis in 1972, calling it "An Empirical Test of a New Theory of Human Need." This model, in which the key attributes are existence, relatedness, and growth, is particularly useful in our attempts to understand both the positive and negative dimensions of the treasure impulse. Although Alderfer developed this model to describe organizations, its applicability is much broader, for it elegantly shows how connections between material needs and more intangible aspects of well-being can be positive as well as negative, depending on the motivational direction (fig. 9). Some kinds of acquisitive impulses and material usage beyond basic needs can have a positive impact on satisfaction, while others do not: much depends on what the motivation and causal mechanism is for the consuming act and what meaning is derived from it.

In general, a certain amount of material wealth is important for individuals to be happy, but wealth is not a sufficient condition for happiness. In making observations like this, researchers are careful in differentiating income from wealth. Income is an economic indicator of monetary flow in households and the economy at large; wealth is an indicator of more lasting financial security. For purposes of understanding material consumption, wealth is a better indicator, since it includes material assets and may account for factors like debt and long-term financial security (which means less worry). For research purposes, however, it is easier to get data on income, so this is what most studies tend to focus on, which can lead to "focusing illusions," according to Nobel laureate Daniel Kahneman: "people exaggerate the contribution of income to happiness because they focus, in part, on conventional achievements when evaluating their life or lives of others."[38] Income alone provides an incomplete picture, so looking at material acquisition in addition helps us better understand the complexities of human well-being.

The mere act of seeking wealth can potentially increase well-being if there is a clear goal in mind, but without clear markers of achievement it can be self-defeating. One of the most important insights from psychological research on human well-being has been that the process of attainment is even more significant than the substantive goal itself. Most researchers

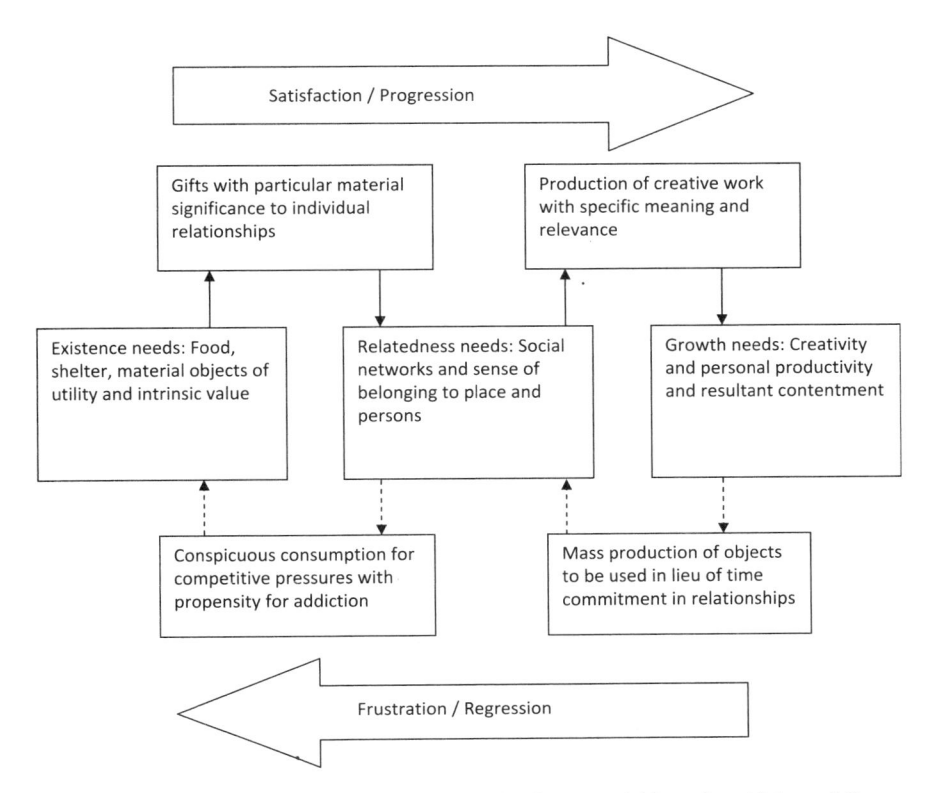

```
                    Satisfaction / Progression  →

    ┌─────────────────────────┐   ┌─────────────────────────┐
    │ Gifts with particular   │   │ Production of creative  │
    │ material significance   │   │ work with specific      │
    │ to individual           │   │ meaning and relevance   │
    │ relationships           │   │                         │
    └─────────────────────────┘   └─────────────────────────┘
              ↑                             ↑
┌──────────────────┐  ┌──────────────────┐  ┌──────────────────┐
│ Existence needs: │  │ Relatedness      │  │ Growth needs:    │
│ Food, shelter,   │  │ needs: Social    │  │ Creativity and   │
│ material objects │  │ networks and     │  │ personal         │
│ of utility and   │  │ sense of         │  │ productivity and │
│ intrinsic value  │  │ belonging to     │  │ resultant        │
│                  │  │ place and        │  │ contentment      │
│                  │  │ persons          │  │                  │
└──────────────────┘  └──────────────────┘  └──────────────────┘
              ↑                             ↑
    ┌─────────────────────────┐   ┌─────────────────────────┐
    │ Conspicuous consumption │   │ Mass production of      │
    │ for competitive         │   │ objects to be used in   │
    │ pressures with          │   │ lieu of time commitment │
    │ propensity for          │   │ in relationships        │
    │ addiction               │   │                         │
    └─────────────────────────┘   └─────────────────────────┘

    ←  Frustration / Regression
```

FIGURE 9. Existence, relatedness, and growth: the material interface (Adapted from a model presented in Clayton P. Alderfer, *Existence, Relatedness, and Growth: Human Needs in Organizational Settings* [New York: Free Press, 1972])

agree that human beings tend to settle very quickly into a state of pleasure and then become bored with their situation—a phenomenon called hedonic adaptation. To remain satisfied, they need to be experiencing change, whether through motivational flow of creative behavior or through a positive quest. Psychologists Philip Brickman and Donald Campbell in 1971 described this constant striving to maintain the same level of satisfaction as a "hedonic treadmill." "The nature of [adaptation] condemns men to live on a hedonic treadmill, to seek new levels of stimulation merely to maintain old levels of subjective pleasure, to never achieve any kind of permanent happiness or satisfaction."[39]

This rather bleak view of human nature has led to many efforts by researchers to refute it. In this chapter we saw how biochemical pathways definitely lead to feelings of joy and the ability of some individuals to feel measurably "happy." Surveys of happiness, despite their many methodological problems, have shown that people are definitely happier in some circumstances and geographies than in others.[40] However, despite the various dissenting views, what has emerged is that positive pursuit is indeed an essential ingredient in well-being. Such pursuit must be anchored in a broader purpose and might not necessarily involve consumption. Building on the treadmill analogy, the individual may not be moving forward by some measures of subjective well-being, but if the goal is framed in terms of physical fitness, they are very much on the right track.

In addition, it is essential to blend subjective well-being with objective criteria of human prosperity, such as health, which usually figures as the most significant factor in well-being according to most surveys.[41] Furthermore, given the transient state of human happiness at any given point, it is highly consequential to consider long-term impacts of consumption at other levels. The connections between consumption, the environment, and human well-being form a subject of increasing academic inquiry. Such scholars as Thomas Princen have been asking us to consider the trade-offs between efficiency of large-scale production-focused economies on one hand and the "logic of sufficiency" on the other. If we consider longer timescales in our decision-making frames, it is more likely that we would act with what Princen terms "ecological rationality."[42] We might begin to consider individual purchasing decisions with greater care. Just as nutrition labels made people more savvy about health-conscious shopping, perhaps awareness of longer-term consumption impacts may lead to greater product differentiation by consumers.

Even with responsible shopping, some ardent environmental critics of consumerism like Juliet Schor would suggest that Americans are "overshopping."[43] If all the little additional trinkets and consumer goods that we buy are not definitively improving our quality of life, then why do we have to go on spending? Part of the answer may be that people have had greater access to credit, often to the detriment of their long-term planning, as exemplified by the financial crisis of 2008. With huge credit availability, Americans have come to view the opportunity for immense wealth as far more reachable than did previous generations. Schor argues that material acquisition may provide the illusion of bridging financial inequality. However, the larger question remains to be addressed about what impact shopping has on the economy as a whole and on livelihoods of those who produce the goods as a trade-off against environmental impact.

After the tragic events of September 11, 2001, American consumers were asked to shop further to help the economy. In a nationally televised speech on September 20, President Bush followed the advice of most economists and urged Americans to show "continued participation and confidence in the American economy."[44] The advice perhaps seemed logical, since it was widely reported that the terrorists were against American "culture," of which materialism was a hallmark. Exhortations to spend and support businesses became increasingly vocal, even as war expenditures increased. If spending would boost company incomes and potentially create more livelihoods, then shopping could be seen as leading to economic growth, so a moral argument could be made for it.[45] However, this approach was in stark contrast to the World War II years, when in times of uncertainty and war spending consumers were asked to conserve. In those days, the famed public intellectual Lewis Mumford wrote of the need for reducing consumption as an act of patriotism. In his view, "The economy of sacrifice turns the economy of comfort upside down."[46] Such conflicting advice within a time span of less than a century may lead us to ponder. Had the economic system changed so much that spending in times of crisis had gone from a vice to a virtue?

Equally strident were calls from the political left to conserve on the basis of environmental harm, given our economic dependence on foreign fuels. The world turned its attention to those regions that were producing the fuels to meet our appetites. Much might be learned from attempting to understand their struggles with achieving affluence. Such countries as Nigeria, Angola, Congo, and Venezuela, having paid for materialism through their abundant supplies of extractive industries, were not faring too well either. Had our cravings for material treasures "cursed" them in some bewildering way as well? The answer may surprise many who are used to a monochromatic vision of resources as a bane or a blessing.

5

CURING THE RESOURCE CURSE:
MINERALS AND GLOBAL DEVELOPMENT

> The iterating of these lines brings gold
> The framing of this circle on the ground
> Brings whirlwinds, tempests, thunder and lightning
> —Christopher Marlowe, *Dr. Faustus*

Most international flights to Kinshasa, the capital of the Democratic Republic of Congo (DRC), arrive late at night, and passengers anticipating a view of this city inhabited by more than 7 million people are usually disappointed. There are no expansive vistas of the mighty Congo River, which skirts the city's esplanade, nor can one see any monuments or structures that would remind us that this is one of Africa's largest metropolises. A comparable city in other parts of the world would sparkle with city lights in the night sky, but Kinshasa's downtown offers a remarkably dim glow of an occasional security lamp. The ambulatory headlights of a few old cars can be seen snailing around some curvaceous streets, but much of the city remains engulfed in darkness. As of 2008, only 6 percent of the DRC's population has access to electricity, the lowest rate in Africa. Out of this national percentage, 35 percent of the urban areas are electrified (accounting for 22 percent of the country's population), while only 1 percent of the rural areas have electricity (where 78 percent of the people live).[1] The low rate of electrification, despite one of the largest hydroelectric potentials in the world, is only one indicator of the lack of development in a country that boasts some of the world's largest reserves of minerals. Gold, copper, cobalt, and diamonds are all abundantly found in this massive, richly vegetated country one-fourth the size of the continental United States. Yet much of this wealth has been extracted without any visible impact on the country's economy.

Traveling around the DRC, one finds reminders of the country's latent wealth in strange ways. Kinshasa has its share of tall buildings, most of which are now abandoned or in varying stages of disrepair. The city's Stade des Martyrs, which hosted the famed "rumble in the jungle" boxing match between Muhammad Ali and George Foreman in 1974, boasts a capacity of one hundred thousand but rarely sees many crowds these days for international events. The University of Kinshasa, supported during the Cold War with American financial assistance as the largest center of higher learning in Africa, has been looted of most of its library books. The university still has more than twenty-six thousand students, but for much of the academic year faculty are on strike or the campus is closed because of security concerns. Amazingly enough, the university has a small experimental nuclear reactor that was built during the Cold War as a compensatory gesture from the United States for Congo's supply of uranium for U.S. weapons. This is supposed to be the most secure site in the country, but during my visit there in 2002, it seemed deserted of much security, and the thought of enriched uranium in this troubled land eerily reminded me of Joseph Conrad's portrayal of the Congo in *Heart of Darkness*.

In the northern jungles of Congo bordering the Central African Republic are the ruins of a city called Gbadolite, another sinister reminder of how the country's wealth was wasted within generational memory. While much of the country languished in abject poverty, President Mobutu Sese Seko, who ruled the country for almost four decades (1961–1997), focused on developing this remote corner of the country from where he originally came to power. An airport was built with a runway to accommodate a Concorde and a terminal building embellished with choicest European frescoes. The town had clean water and reliable electricity during its heyday in the 1970s. One of the finest high schools in Africa was built to educate the population of the Equateur province that had supported the regime. The hospital facilities and roads were comparable to those in a developed country. Grand palaces were built to accommodate the ruling elite, and many who saw its glory days still recall the city as a Versailles of the Jungle. Yet the prominence of Gbadolite was short-lived, as such asymmetries of wealth were clearly unsustainable. The Mobutu regime finally fell to rebels in 1997, and the despot who had looted over $5 billion in wealth from his land was forced into exile in Morocco.[2] More than a decade later, peace has still not returned to Congo, and the country remains trapped in one of the most excruciating civil wars in history, one that has taken more than 3.9 million lives since Mobutu's exile.[3]

What went wrong with the DRC's development trajectory? Was Mobutu to blame for all of the DRC's woes? Even if he was the proximate cause, we are still confronted with the question of how he was able to assume power and retain it for thirty-seven years. How did a country with so much potential languish and atrophy into what Paul Collier has called "the bottom billion"? Economists, political scientists, and sociologists have all puzzled over how a country rich in resources can be so abysmally underdeveloped and stricken by conflict. Some have termed the condition a "resource curse," which should make mineral extraction for new states taboo. Is a scramble for resources to blame for conflict, or are incipient inequalities and economic injustice the primary cause? Poetic alliterations such as "greed versus grievance" or "the paradox of plenty" have animated the literature and caught the public's imagination.

Is there a connection between the kind of resource being extracted and the propensity for underdevelopment? What is the likelihood of violent conflict developing in these conditions? For example, while both the DRC and its copper-rich neighbor Zambia had minimal development investment from mineral revenues, Zambia was able to avoid violent conflict for much of its history. Farther south, Botswana exemplified quite the opposite: mineral wealth transformed a struggling economy into one of the most stable and educated democracies in Africa.

How might one account for intervening variables that could explain these differences? Might we compare the brutality of Belgian colonialism under King Leopold in Congo with the relatively more "civil" colonialism of the British in Zambia and Botswana?[4] Was leadership a key factor in the divergent trajectories in the postcolonial era? The rise to power of the incorruptible Sir Seretse Khama in Botswana or the relatively benevolent autocrat Kenneth Kaunda in Zambia may be contrasted with the unfulfilled promise of Patrice Lumumba, who was assassinated with Belgian complicity in Congo soon after independence. Ethnic homogeneity could perhaps be another explanatory variable—the demographic dominance of one ethnicity in Botswana (the Setswana) versus the enormous diversity of tribal populations in the vast geographic space of the DRC.[5] We could also posit alternative hypotheses about the kind of mining that was being undertaken in one country versus the other—large-scale mining in Botswana and Zambia, versus small-scale lootable diamond mining in Congo and later in Sierra Leone.[6] Were resources doomed to become a cause for avaricious foreign intervention and instability? Were the "great games" of international relations largely played with mineral pawns and often an impetus for foreign misadventures, especially in oil-rich countries of the Middle East? These are all exciting questions for social scientists to address and have been very attractive for the public at large as well. But because of the complexity of social processes, causality is hard to extract from these examples. Many researchers have decided to resort instead to quantitative analysis to see what patterns may be found in the economic and political fortunes of resource-rich countries.[7]

The Richness of Numbers?

Scientific enterprise in the past century has been dominated by numbers, because they appear to provide an objective basis for analysis. When one is grappling with complex global problems, such as resource dependence and poverty alleviation, numbers can offer a shield against political innuendo. Even disciplines that had historically used qualitative methods, such as sociology and political science, are now dominated by numerical analyses. The supremacy of numbers was enunciated by the British physicist and father of modern temperature measurements, Lord Kelvin, as follows: "When you can measure what you are speaking about, and express it in numbers, you know something about it; but when you cannot measure it, when you cannot express it in numbers, your knowledge is of a meager and unsatisfactory kind. It may be the beginning of knowledge, but you have scarcely, in your thoughts, advanced to the stage of science."[8]

At one level this assertion makes sense, because facts are often articulated in numbers—demographic indicators of communities, criteria for environmental impact, prices of commodities, and development indicators inherently need measurement for cogent comparison. However, when numbers are used in more abstract forms to argue for causality, then we are in a different domain altogether. As noted by science historian Theodore Porter, Kelvin's words were meant as a critique of "nihilism" rather than a blanket support of numerical causality research. In his notable book *Trust in Numbers: The Pursuit of Objectivity in Science and Public Life,* Porter provides an appropriately measured critique of the explanatory power that numerical analyses can provide in complex global problems.

Of particular interest in understanding the dynamics of the "resource curse" is the use of regression as a statistical technique. Since its development by the French mathematician Adriene-Marie Legendre, the technique gained rapid prominence in all scientific disciplines. Francis Galton, a cousin of Charles Darwin, first used the term "regression" to describe the mathematical technique he applied in studying the progeny of gifted adults.[9] Regression is an elegant technique that allows for independent variables to be compared in relation with dependent variables in a classic scheme of hypothetical testing. However, it is also a deceptively powerful technique because a positive finding is by no means definitive and is often very hard to disprove unless one has a clear understanding of intervening variables. For example, let us consider one study, titled "Was the Wealth of Nations Determined in 1000 B.C.?"[10] As the research question in the title suggests, the study, conducted by a highly reputable group of researchers, used regressions to consider if the development and adoption of ancient technologies led to certain societies succeeding more than others. Although this is an interesting question to consider, one wonders how an affirmative or negative response to this question might affect contemporary policy. Given the enormous level of variability in technology transfer across regions and current processes of globalization, might any finding of such a study be of much use? Might certain suggestions from regression research of this kind potentially be misused because of the allure of numbers?

Statistician David Freedman, who used quantitative methods for much of his career, raised concerns about regression modeling when he stated in an article published by the journal *Sociological Methodology* in 1991: "I do not think that regression can carry much of the burden in a causal argument. Nor do regression equations, by themselves, give much help in controlling for confounding variables." Despite his use of regression for much of his career, Freedman was concerned about the increasing applicability of the technique as a touchstone for rigor and a driver of policy.[11]

Regression analysis of large sample data sets became increasingly important to the scholarship on environment and security beginning in the late 1990s. Several international organizations, for example, the United Nations Environment Programme, and a range of researchers from the fields of security studies, development economics, and international relations began to examine the role of environmental factors such as natural resources in their investigations of violent conflict and underdevelopment. Minerals were an interesting variable to consider in the causality for underdevelopment because of the counterintuitive appeal of posting "the paradox of plenty," and also because they had been stigmatized as a problem sector by many earlier economists. Even Adam Smith, considered by many as the progenitor of modern economic thought, had cautioned governments about mining projects in his monumental treatise *The Wealth of Nations:* "Projects of mining, instead of replacing capital employed in them, together with ordinary profits of stock, commonly absorb both capital and stock. They are the projects, therefore, to which of all others a prudent law-giver, who desired to increase the capital of his nation, would least choose to give any extraordinary encouragement."[12]

Even in the realm of natural science, resource-curse hypotheses flourished. The German botanist Matthias Jakob Schleiden, who is credited for some pioneering work on understanding cell biology, also wrote a book titled *Das Salz* in 1875, which described how there was a direct correlation between salt taxes and tyrants.[13]

Most notable among the originators of the resource-curse hypothesis in current social science discourse were British economists Richard Auty and Paul Collier, who consulted for various international organizations, as well as notable public intellectual and United Nations adviser Jeffrey Sachs.[14] Using the ratio of the financial value of primary commodity exports over gross domestic product (GDP) as a measure of natural resource abundance, Paul Collier and Anke Hoeffler found a significant relationship with violent conflict. More specifically, they found that the risk of civil war onset was greatest when the GDP share of primary commodity exports for a country was around 35 percent.[15] Some subsequent cross-national statistical analyses found that natural resource abundance in a particular sector (such as oil or minerals) was a cause of violent conflict.[16]

Several environmental and human rights groups, including Oxfam, Global Witness, Earthworks, and the Worldwatch Institute, have supported this perspective in their research and analysis on cases of conflict over diamonds.[17] Some researchers have found that the implementation of peace agreements is likely to be more difficult in countries abundant in natural resources, which can pose difficulties in postwar peace-building efforts.[18] The list of studies is almost endless, with each scholar trying to find a way to capture the complexity of national destiny through ever more complex numerical analyses.

Depending on the assumptions and resolution of the regression, there have been highly divergent findings on conflict linkages with mineral economies as well. For example, James Fearon and colleagues at Stanford University found no such linkage despite including a relatively comparable set of civil wars in their data set and operationalizing resource abundance in the exact same way as Collier and Hoeffler.[19] In a follow-up article, Fearon confirmed that minor departures from the Collier and Hoeffler approach, such as using yearlong data rather than grouping data in five-year intervals or employing a multiple imputation statistical methodology to include cases with missing data, resulted in primary commodity exports no longer being associated with civil war in statistical terms.[20] Subsequently, the results of a rigorous analysis of thirteen most likely case studies yielded little evidence that looting of legal agricultural commodities or the presence of oil were related to funding the startup costs of insurgency movements, and in only one case were "booty futures" in oil revenue used to finance a rebellion.[21]

Proponents of the resource-curse theory postulate that oil abundance is particularly pernicious, not primarily because of its financial value to rebels, but because oil-abundant countries tend to lack strong state bureaucracies. Oil abundance minimizes the need to collect general tax revenues, which is often one of the major functional reasons for building capacious state agencies. The statistical results support the weak-state hypothesis, but many of the measures widely used to operationalize state strength—government observance of contracts and investor perception of expropriation risk—are not particularly compelling proxies. But this argument and supporting evidence of weak state institutions being associated with civil war outbreak is ironically consistent with the ideas and results offered by scarcity researchers that strong political institutions can reduce the likelihood of violent conflict in developing countries. In perhaps the most comprehensive conceptual and statistical work to date, Macartan Humphreys finds support for the weak-state mechanism rather than the rebel financing hypothesis.[22] Conflict initiation is correlated with past resource production (for both oil and diamonds) rather than potential future production (oil stocks). But the findings suggest that natural resource abundance also affects conflict onset independently of state strength. Contrary to previous research, Humphreys discovers that violent conflicts in which natural resources are at stake tend to be shorter in duration and often end in the military defeat of one of the combatants.

The debate between environmental abundance and scarcity approaches has not progressed very far, partly because the scholars in these communities do not engage directly with each other very much. Perhaps even more important, there are clear differences in research style and the philosophy of knowledge between the two perspectives. Indeed, some systemic problems of dependence have been identified in economies that export natural resources because they lead to a stifling of local manufacturing industries. The process begins with a rise in exchange rates due to increasing exports, which in turn makes local manufacturing less attractive and entangles the public sector intimately with business interests. This phenomenon occurred in Holland in the mid-twentieth century when natural gas was discovered, giving rise to the term "Dutch disease."[23] Yet the country was clearly able to overcome any effects that may have occurred through policy innovation, and it is now one of the more prosperous countries in Europe. Economic geologist Graham Davis suggests that countries with resource endowments can reconfigure the windfall revenues in such a way that they can "learn to love the Dutch Disease."[24]

These analyses have in turn been coupled with theories of governmental failure at various levels. Countries dependent on natural resources are sometimes called rentier states—a term from the work of Iranian scholar Hossein Mahdavy that describes petroleum-exporting countries that receive most of their national income from the rent derived from their natural resources being extended for extraction to external clients (primarily export revenues).[25] Rentier states are characterized by a lack of dependence on tax revenues due to their mineral royalties and hence reduced accountability as well. Additionally, the mineral royalties could be spent on either buying political favors or repressing the population. This has led to a series of fatalistic accounts, most notably by Thomas Friedman, whose "first law of petropolitics" suggests that freedom and democracy in oil-exporting states are inversely related to oil prices.[26] Michael Ross, who has spent much of his career studying the resource curse through regressions, has gone so far as to propose that oil is to blame for women's rights violations in the Middle East.[27] Although Ross's analysis may be impeccable in its quantitative methodology, it raises fundamental questions about how much one can suitably infer from regressions that are unable to account for myriad cultural variables that are coincident with geological availability of oil. Anthropologists who have studied Middle Eastern cultures long before the advent of oil would shudder at such an explanation, but the popular media were attracted to this unusual "finding."[28] Exonerating cultural factors and the nuances of development in such stark causality can clearly be quite dangerous from a policy perspective. Yet the furor and fervor of resource-curse determinism prevails, particularly with reference to fossil fuels.

Oil and Turmoil?

As rebel troops rolled into N'djamena, the capital of the African nation of Chad in March 2008, commentators were ready to blame it all on the country's oil. To many, it was the resource curse in action: oil-rich countries being driven to civil strife by regional avarice and economic distortions created by a sudden influx of wealth. The headline on CNN immediately read: "Oil fuels ethnic violence in Chad."[29] Environmental groups and human rights activists felt vindicated that their campaigns against an oil pipeline that had been built with World Bank assistance through Chad and Cameroon would now be taken more seriously. Given the backdrop of the war in Iraq, which has been branded an "oil war" by many, there was a general presumption that the ultimate cause for conflict in this remote part of Africa was that same loathsome liquid.[30]

The connection between oil and conflict has been made as far back as the earliest industrial uses of the fuel. Soon after the end of World War I, the

French oil executive Henry Berenger in a historic dinner speech alongside the distinguished British diplomat Lord Curzon argued that "as oil had been the blood of war, so it would be the blood of peace."[31] If oil was part of the problem it would perhaps be part of the solution as well. Yet we need to consider the complexity of conflicts in regions like Chad far more carefully before assuming linear causality. First, the conflict in Chad and other "new oil" countries in Africa may predate the discovery of oil in those regions, and thus the underlying ethnic rifts may be a more profound determinant. Civil war in Chad predates the discovery of oil by at least two decades, and the country had never been a beacon of prosperity.[32]

Just as the conflict in Chad was unfolding, events in a neighboring African country were casting doubts on the oily connection with conflict. One of the most stable states in the region was descending into chaos without any incriminating oil to blame. Kenya's postelection chaos in spring 2008 revealed how visceral ethnic tensions can escalate into violence even in nascent democracies with not a drop of oil to fuel the rage. There are so many intervening variables in conflict causality in Africa that blaming oil as the dominant force is not empirically justifiable, no matter what selective regression analysis may suggest. That being said, we must recognize that a sudden influx of cash without appropriate planning and national asymmetries can be a recipe for disaster with any commodity. Extractive industries are a kind of windfall development similar to the establishment of a casino in an impoverished neighborhood, ushering in a sudden influx of wealth to a community. In order for such a windfall development to be successful in the long run, it must be coupled with some other development strategies that utilize the revenues from oil and minimize its environmental impact.

However, with the growing influence of globalization on national policies, some of the fears of resource dependency in Africa and its connection to corruption may be assuaged. Let us take the example of Equatorial Guinea, which has been a languishing dictatorship since its independence from Spain in 1968. After the discovery of oil in the mid 1990s, the international community became more engaged with this tiny country. The United States reopened its embassy in Malabo in 2003, and the State Department asserts that U.S. "intervention has resulted in positive developments," such as an office to monitor the human rights situation in the country. The viability of such a mechanism as a means of initiating change in Equatorial Guinea was tested by a scandal involving the alleged siphoning of funds from oil revenues to an account held by President Teodoro Obiang's family at Riggs Bank in Washington. There were direct linkages made to acquisition of property in Washington suburbs from these accounts. This led to a U.S. Senate hearing on the issue and an investigation by the Office of the Comptroller of Currency in 2004. None of this would have happened if the salience of Equatorial Guinea had not been brought to the world's attention by oil. Yet the onus for exerting such influence still lies with the international community.[33]

At the same time, the regulatory capacity of some African governments over oil activities has grown. Some governments have gained expertise in the technical matters of the oil business, improving their capacity to negotiate concession contracts and regulate social and environmental issues. For example, Angolan authorities fined Chevron-Texaco $2 million because of an oil spill in the sea in 2002.[34] Even though oil companies have a history of oil spills and pollution in the region, this was the first time one was fined because of environmental degradation in Africa. Chevron-Texaco also compensated local fishermen for losses in their incomes. This is a departure from the days when these companies could operate with relative impunity in nearby Nigeria.

Much of the alarm regarding oil in Africa stems from the admittedly horrifying experiences of the continent's largest oil producer and most populous country—Nigeria. For more than fifty years, oil has been extracted from the southern part of the country, in the Niger delta, yet many of the most oil-rich communities are utterly impoverished and may not even have cooking oil available for their homes. How did those communities that bore the brunt of oil extraction's occupational and environmental hazards not benefit from the bonanza? The answer largely lies in a misconceived attempt at national wealth allocation that befell Nigeria in the postcolonial era. A country already fractured by ethnic and religious differences tried to extract wealth from one location and invest whatever escaped the clutches of corruption through a process of demographic prioritization. Urban areas were thus given priority, and lavish expenditure was incurred in building a new planned capital, Abuja, in the heart of the country. A diabolical confluence of errant planning, poor leadership, artificially crafted national geographies, and apathy from multinational corporate investors pushed the country into chaos. Much to Nigeria's misfortune, it received investment from foreign companies at a time of minimal corporate accountability and transparency. There is little doubt that the extractive sector has historically had a strong linkage to corruption. In one study conducted by the Gallup organization, the supply side of international bribery was measured in a Bribe Payer's Index. Oil, gas, and mining had the second most negative ranking in this index, after the arms industry.[35]

The question to consider in such eventualities is whether the situation might have been any better if oil had not been discovered. Would alternative sources of income have been available to sustain the population and conflicts been averted? In countries like Saudi Arabia, Malaysia, or even Venezuela, some opportunity-cost analysis of alternative sectors would probably suggest that oil in aggregate made things better in development indicators. But the Niger delta is a region where it is safe to say that oil actually diminished the development prospects for that part of the country. The main reason was that the alternative livelihoods of the communities were themselves impaired. Agriculture, fishing, and the potential for tourism were all harmed by the negligent way that oil was extracted. Clearly, this is a stark omen for countries to consider. However, once these systemic roots of why oil led to turmoil in the Nigerian case are appropriately considered, we should apply those lessons in future development efforts rather than dispensing with them altogether.[36]

As peace returns to the streets of Chad, the eye of the international community should remain on how the oil revenues are managed and how the country ultimately plans for a post-oil economy. The elaborate system for revenue transparency that the World Bank had set up for Chad's oil broke down in September 2008. This failure of a promising system of financial accountability occurred because international institutions, led by such large investor states as China, were not willing to enforce the arrangements with sanctions or other punitive instruments on the country.

Despite its tortured history and its ultimate demise as a fuel, oil must not be summarily dismissed as a cause of turmoil in Africa. Rather it should be considered as a resource that needs to be managed with effective development planning for countries with few other alternatives, such as Chad. Other exogenous questions about whether oil should be used by society due to climate change are usually moot in the context of these particular decisions, since we are unable to transition immediately to a non-oil economy and will need at least some quantity of oil while alternatives are developed. No doubt, oil development in this context should not be used as an excuse for complacency or delaying adoption of alternatives.

Additional safeguards are needed for oil more than other commodities because of global dependence on the mineral for energy, its external impact on climate, and because of the checkered past of its usage. However, avaricious scrambles for resources are by no means confined to oil, and we would be going down a very slippery slope if extraction of a resource is summarily discouraged because of poor planning. Human demand for many unlikely materials, even edibles, can prove to be just as contentious as oil.

Of Fungi and Ice Cream

Far from any oil wells, near the forested reserves that surround one of the great calderas of North America, there is another scramble for resources under way. The rush is largely invisible to most visitors to Crater Lake National Park, unless you hike around it and encounter some signs that forbid the extraction of an item that is supposed to smell like a mixture of "red hots and smelly socks."[37] This unappealing commodity that has attracted migrant laborers from as far off as Cambodia to extract is a lowly mushroom. The fungal kingdom, which is crowned with mushroom species, hosts some of the more bewildering kinds of organisms.

A renewable and often unwanted entity such as a mushroom can be the subject of a scramble for extraction when the market conditions are right. In an ethnographic article for the *New Yorker,* veteran nature writer Burkhard Bilger documented an extent of criminal activity and shadow marketing that one usually associates with minerals rather than mushrooms. In the words of the owner of a campsite for itinerant mushroom hunters in this region, interviewed by Bilger: "It gets pretty wild. You get the drugs, the prostitution, the robbing at gunpoint, the gangbanging crap."[38]

The fungal hunters in this part of the northwestern United States are looking for a particularly marketable variety known as Matsutake, which is prized in Japan for its flavor and reputed properties in stimulating male virility. The stout-stemmed, knobby mushroom's shape and size readily betray the rationale for this age-old belief in Japan, where the prized fungus is flown daily from Seattle and Vancouver in gel-chilled Styrofoam boxes. A few pounds of good-quality mushrooms could earn a picker several hundred dollars, and the price would be magnified several-fold, depending on the domestic seasonal availability in Japan. The price is highly capricious, even more so than commodity metals, because of the unpredictability of a harvest that depends on pest infestations, heat waves, and nutrient availability. Due to a bad domestic harvest year in Japan, the wholesale market price in 2007 was as much as $200 per pound, with particularly appealing individual mushrooms fetching as much as $150 per stem.[39]

One might wonder why such a prized commodity is not cultivated artificially at a larger scale to reduce cost and the potential for such a shadow market. As with chanterelles and porcini, the Matsutake mushrooms defy all efforts at cultivation. Despite efforts by mycologists from Japan to Sweden to find the perfect mixture of humidity, temperature, and nutrients, transplanted organisms wither away in days. Clearly, some mysterious mixture of factors in temperate old-growth coniferous forests keeps these coveted treasures away from human shortcuts. Their natural availability makes them quite similar to minerals, and one may ponder whether this attribute makes them more susceptible to the kind of "rush" behavior as well as the negative influences of rampant extraction.

Yet if we consider other commodities that may be cultivated, one can encounter similar patterns of influence. Consider the most favored flavor in ice cream, found as a bean in the wild but also cultivated or synthesized artificially. The Vanilla Coast of Madagascar provides some useful comparisons in this regard. Long before a sapphire rush took over the economy of the southwestern part of the island nation, the northeastern corner of the country was well regarded as an ideal locale for growing the delicately aromatic bean that graces creameries and bakeries all over the world. Walking around the capital, Antananarivo, one encounters scores of young hawkers trying to sell bundles of the bean wrapped in paper at prices that range with the gullibility of the buyer.

The bean became a national economic symbol in Madagascar, with the World Bank conducting a major study in 2006 to consider national policies on vanilla and its value in poverty alleviation. There is considerable potential for employment, because vanilla cultivation is highly labor intensive. Crop husbandry requires 100 working days per acre during the first year, and about 180 during the four to eight years when plants reach maturity. Weeding and pruning are essential and must be accompanied by hand pollination: each flower on the vine has to be pollinated individually and at different times. The curing process involves dipping the beans in near-boiling water, which starts an enzymatic reaction by alternate heating and "sweating," a method of exposing the beans to sunlight. The process is repeated ten to twenty times before the beans are left to dry outdoors for two to three months, after which they possess a uniformly dark color and release their coveted fragrance.[40] Yet despite this potential for livelihoods, exotic appeal for a tourist sector, and the allure of renewability through cultivation, the vanilla industry has never really taken off as a source of development for Madagascar despite its prevalence for several decades. Errors in governance can be blamed here as well, even though the structural vulnerabilities of the mineral sector do not apply.

Just as the vanilla market was plateauing, a phenomenal sapphire rush engulfed Madagascar, attracting the attention of various development donors. The Projet de Gouvernance des Ressources Minérales (Project of Governance of Mineral Resources, or PGRM) is a particularly interesting example of how a transformation of the mineral sector might also avoid the resource curse and provide some impetus to other sectors. Established in 2003, it provides a new model for development assistance in a traditionally nonrenewable resource sector. The implementation of this project in an ecologically sensitive country with some of the world's most unique biodiversity is particularly appealing. The project is supported by, among others, France, the United States, and South Africa as well as the World Bank. The stated goals of the PGRM include assisting the government in implementing a strategy to accelerate sustainable development of the mining sector and thereby contribute to reducing poverty in Madagascar. One focus is on supporting the small-scale mining sector and handicrafts industries.

The PGRM effort sets out from a scientific base, enhancing the technical and geological skills to find the best mining areas. For instance, a computerized system is in use to interpret and exploit certain geoscientific data. On the other hand, the program boosts business opportunities, for instance, through skill training. Among other things, an Institute of Gemology has been established. At subsidized rates, it teaches Malagasy people the use of state-of-the-art equipment. Scholarships are available, and students from other African countries are increasingly enrolling in courses. Today, some miners have moved up the supply chain in search of a more sustainable livelihood. If Madagascar succeeds in developing a jewelry manufacturing sector, processing and handicraft skills gained at the institute should be of lasting value, even once Madagascar's minerals are depleted. In this regard, it is useful to consider the cases of Chantaburi in Thailand and Surat in India, where specialized skills in gem treatment and cutting made them centers of commerce even after mining stopped in the region. In 2005, *The Economist* estimated that twenty-five pounds of gems were smuggled from Madagascar to Thailand every week for processing—to the economic detriment of Madagascar. Surat, India, has acquired a dominant market share in jewelry manufacturing, even though neither the gold nor the gems are any longer from India itself. Heat treatment of colored gems to enhance their color has been perfected by the merchants of Chantaburi, which makes it a haven for gem traders to this day, long after the mines nearby were depleted. These examples show that it makes sense to invest in a manufacturing sector that ties in well with local skills and market interests.[41]

The opportunities for transforming a linear primary sector like mining into a more productive and diverse enterprise are thus quite plausible. In the same way, a sterile vanilla industry can also be transformed into one that hosts associated products (such as branded chocolates, ice creams, or essences) that can invigorate the economy even further. Governance mechanisms need to be instituted whether one is dealing with noncultivable mushrooms, cultivable vanilla, or nonrenewable sapphires. Let us not forget that some of the most horrific historical episodes of colonialism and the slave trade were motivated not by minerals but by a seemingly harmless need-based carbohydrate commodity like sugar. From Fiji to Haiti, colonial powers scoured the earth to find fertile lands and labor to harvest cane, cashews, rubber, and timber of all sorts. Renewable resources were just as culpable in such nefarious enterprises as nonrenewable ones. Even some of the contemporary conflicts that have gained much disrepute of late, such as the Liberian civil war, were fueled just as much by illegal timber as by diamonds.[42] Each commodity will have its own structural needs from government to be effectively managed, and the resourceful leader will strive to make the most with the least amount of extractive inputs.

Lifting the Curse

Coming back to Congo, diamonds, gold, and a rare mineral called coltan (short for columbium and tantalum) were blamed for the country's woes by the United Nations Security Council as well as numerous nongovernmental organizations. Indeed, the revenues from mining may have contributed to rebel movements in the east of the country. But let us not forget that neighboring Rwanda had been scourged by civil strife and a horrendous genocide a few years earlier without a clearly defined scramble for minerals. Underlying ethnic tensions coupled with massive deprivation and perceived social injustice may well be the root causes of conflict. What the people of Congo need now is not more doom and gloom from the resource-curse theorists but rather an effective means of managing and monitoring the flow of their mineral revenues so that they cannot be misused.

Most of the world's major economic and educational powers have depended at some point in their development path on the exploitation of their natural resources—the United States, Canada, Norway, and Australia being foremost among them. In many cases extractive industries have been necessary and desirable ingredients in the development trajectory. Yet another questioning analysis of resource determinism was published in May 2008 by the journal *Science,* which stated that "the last word in the resource curse debate is far from having been spoken; but economic advisors should be aware that natural resources do not necessarily spell doom for development. Instead, their exploitation can be a valuable part of a sustainable development strategy."[43]

The key challenge, then, is to find the best structural mechanisms to make resources work for communities and for society as a whole. Finding the right mix of market incentives, regulation, planning mechanisms, and community vigilance is essential in making effective use of resources for development and for ensuring that the life-support systems on which we depend are conserved. Through trial and error and with an evolution of innovative programs worldwide, we are beginning to see some progress in this regard. We are now beginning to see scholarly publications with titles such as *Escaping the Resource Curse,* suggesting that policy analysts might be turning the curve on the fatalistic rhetoric of yesterday.[44]

Such paths to development must, of course, be tempered with environmental performance, economic diversification, and a respect for the safety of those who toil to retrieve these resources for us. There has been considerable research on the economic performance of resource-rich countries by academics as well as by international financial institutions. Unfortunately, such studies do not consider the full spectrum of alternative development paths and cannot account for the myriad intervening variables that may lead to poor economic performance or conflict. Some of the most remarkable success stories of development, such as Botswana, Malaysia, and Chile, have been mineral producers, and must not be dismissed as outliers. There are even cases where mineral resources have given a community an advantage in escaping some of the most pernicious aspects of colonialism, as exemplified by the Royal Bafokeng Nation in South Africa. This community of around three hundred thousand has maintained relative autonomy within the country and is economically self-reliant owing to a huge reserve of platinum on its land. The community was initially assisted in maintaining its land and resources with the help of Lutheran missionaries who kept land titles on behalf of the community during apartheid. One of the world's largest platinum reserves was found on their land in 1921, and a protracted struggle to assert rights over this reserve ensued. Ultimately, the community won its legal battles, and the platinum has been an enormously beneficial source of income and in setting precedents for self-determination in South Africa.[45]

As international norms for accountability evolve, the fear that resource endowments will be misused economically is abating. In 2007, forty-one financial institutions agreed to a revised version of the "Equator Principles" for ensuring social and environmental consciousness in their investment decisions. The British and Norwegian governments have been leading the Extractive Industries Transparency Initiative, which aims to recruit governments in having constructive action plans for resource revenues. The Kimberley Process for diamond certification and the partnership between jewelry manufacturers and such groups as Earthworks, Revenue Watch Institute, and Oxfam on "clean gold usage" are promising signs of constructive engagement.[46] Such efforts are far too easily labeled as "greenwash," and deserve more credit and time to mature and are likely to be successful as they gain traction. Consider the vast expanse of diamond production and consumption patterns across the world, which need to be analyzed in enforcing such governance structures (fig. 10). Curing the social ailments of resource dependence is possible but requires us to consider clear mechanisms and be ready and willing to innovate as demanded by the times. Some of the peculiarities of large-scale investments in the mineral sector, such as a lengthy exploration phase for project development and captive capital once the mine is built, have to be taken into account as well.

Diamond discoveries in the United States were largely confined to the southern state of Arkansas, which still boasts the Crater of Diamonds State Park where many tourist treasure seekers can still be found sifting through gravel in search of a prized gem. Although Hillary Clinton wears a four-carat canary-yellow diamond ring from this mine, Arkansas never became resource dependent on diamonds or any other mineral. The state's claim to fortune and fame came instead from the retail sector, thanks to a man named Samuel Walton, who grew the world's largest corporation, Wal-Mart, beginning in the small town of Bentonville. The same kind of scrutiny about supply chain accountability, labor practices against economies of scale, and market dominance have come up with this corporation as well—remarkably similar to the narratives against De Beers and other diamond companies. Activists raised concerns about why the company's presence had not provided greater benefits to the state, which is still ranked forty-eighth in median household income. Clearly such concerns about resource flows from corporate enterprise transcend any particular industrial sector.

Another state that became a textbook case of citizen accountability regarding the management of resource revenues is Alaska. The state had to

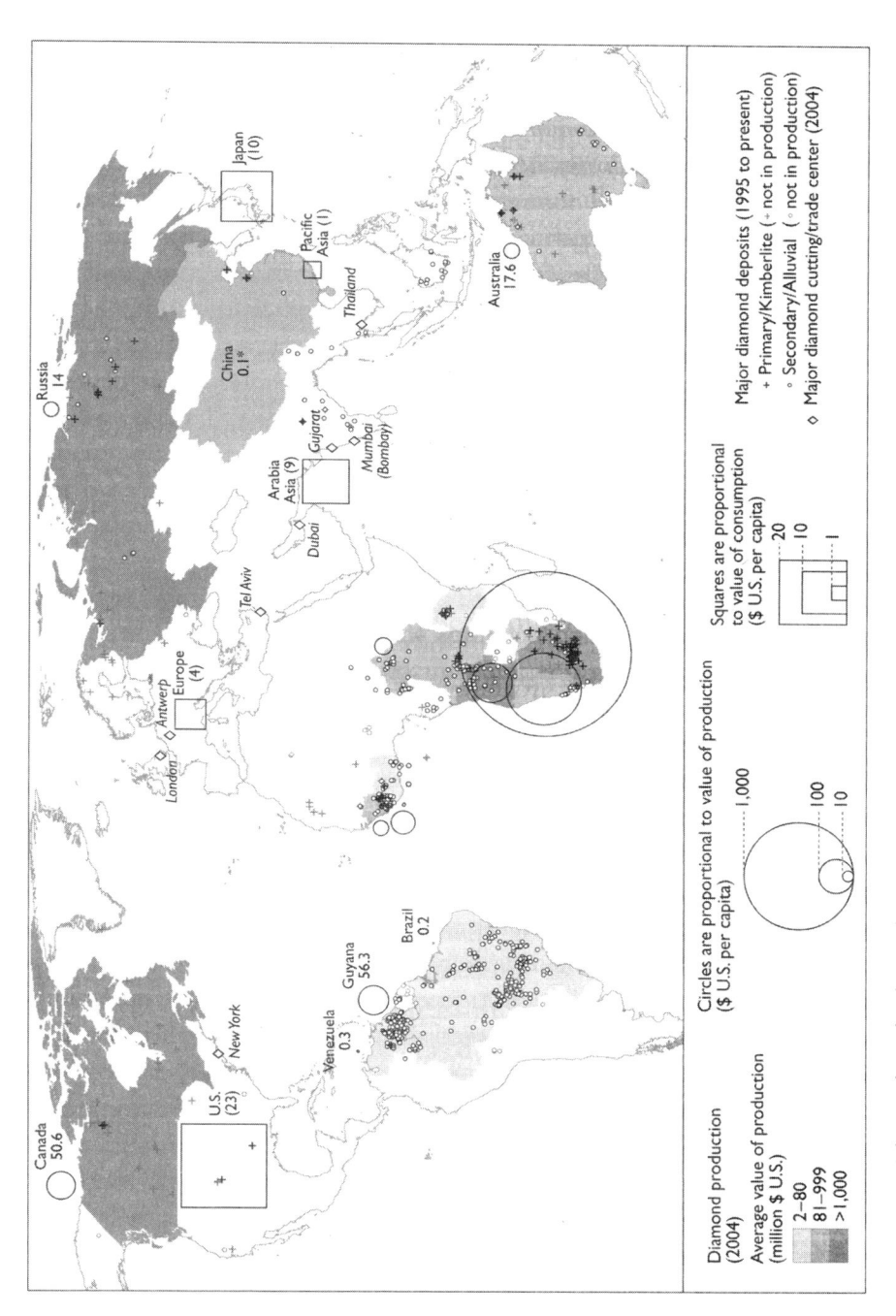

FIGURE 10. Diamond production and consumption worldwide (Philippe LeBillon, "Diamond Wars? Conflict Diamonds and Geographies of Resource Wars," *Annals of the Association of American Geographers* 98 [2008], 345–372)

change with the times in remarkable ways ever since it was bought from Russia by the United States for a paltry sum of $7.2 million in 1867, which amounted to around 1.9 cents per acre (the purchase price today would be around $105 million in current dollars). Despite its arctic climate, which might be a discouraging factor for migration, residence in Alaska is much sought after and considerably difficult to achieve because of the state's oil wealth. Each and every resident has a stake in this wealth because of an innovative system created in 1976 to ensure long-term stability of the state's economy. Funded through royalty payments to the state from oil extraction, the fund started with only around $700,000 and is now worth more than $40 billion. Every year the citizens of Alaska receive a dividend payment from the fund—about $1,700 each in 2007. The ultimate goal of the fund is to provide for an investment base for economic diversification of Alaska's economy once the oil stops flowing. Similar arrangements have been made by the Norwegian government, which now has a petroleum fund estimated at around $200 billion. However, such funds require good management, transparency, and delivery. Venezuela set up a similar fund called the Macroeconomic Stabilization Investment Fund, but because of executive interference over the years, it has been relatively ineffective.[47]

The texture of corporate ownership of mineral companies can also play a major role in determining whether wealth is used beneficially or not. Saudi Arabia realized in the 1970s that it needed to transition from private ownership to state ownership with foreign investment in technical expertise. Through carefully calibrated negotiations between the Americans and the Saudis, ownership was gradually transferred from American companies to the Saudi government in 1980. Foreign capital usually plays an important role earlier in the exploration and development cycle, but eventually some level of state ownership is usually beneficial to ensure appropriate wealth transfer to local populations.[48]

The transformation of natural resource wealth to productive enterprise and livelihoods is an ongoing challenge for humanity and one that can entice us into rash behavior but can also catalyze lasting development. Appreciating the ephemeral state of mineral resources in their natural form is essential to avoid the eventuality of a resource curse. Yet the very evanescence of an ore deposit may lead some to rush and more ravenously exploit the resource while prices are high and each extractor asks that fundamentally selfish question: "How long will the good times last?" This is where planning policies have to moderate the speed and guide the trajectory of resource exploitation. Among the most compelling and natural arguments in this regard is to consider the environmental toll of the mining and drilling processes and whether the life-giving systems that support human societies are being irreparably damaged by extractive excitement.

6

THE SPOILS OF THE EARTH: THE ECOLOGICAL TOLL OF EXTRACTIVE EXCITEMENT

> Summer had changed beyond recognition the winter's trail. Dams of cribwork filled with stones, flumes, and sluice boxes lay across our paths; heaps of tailings glistened in the sunlight beside yawning holes with windlasses tumbled in; cabins were deserted—the whole creek, wherever work had been done was ripped and gutted. Nothing but flood and fire is so ruthless as the miner.
>
> —Tappan Adney, *Klondike Stampede*, 1898

"The Last Frontier" is emblazoned on the license plates of Alaska, which has the largest acreage of truly pristine wilderness in the United States. For Americans, the state has always been a perplexing paradox. It is the largest in size but smallest in population density; it has the coldest climate but the most active system of fiery volcanoes; and it has been the largest source of a nonrenewable resources like oil but also the largest source of renewable resources, such as timber and fish. The state is home to the greatest percentage of native people in the country, with over 15 percent of the population of indigenous Alaskan lineage, but unlike much of the lower forty-eight states, there were no treaties signed between settlers and locals. The natural resource rights were thus negotiated much later through the Alaskan Native Claims Settlement Act in 1971, in which a unique system of twelve native corporations and two hundred village corporations was established. Unlike the reservation system in other parts of America, which relied on government intervention for economic development, the Alaskan corporations were modeled after a free-market approach to development.

The results have been mixed in terms of economic performance between the various tribal populations, but a common theme of ecological stress has emerged across the state, raising some essential questions about how we perceive the environment in a resource-rich wilderness.

Adding to the list of Alaska's extremes is another unlikely statistic. According to the Environmental Protection Agency's Toxic Release Inventory, Alaska emits the largest amount of toxins into the air, land, and water of any state in the union. This dubious distinction has been held by the state for six years in a row since 2001, and raises alarms when mentioned to Alaskans as well. What could possibly be so polluting? The usual suspect is the North Slope's oil, but this is in fact extracted without much incident and piped fifteen hundred miles to the port of Valdez. Problems may arise once the oil is on the tankers, but those incidents have been relatively rare since the devastating *Exxon Valdez* disaster in 1989, which spilled nearly 11 million gallons of crude into Prince William Sound.[1]

Alaska's top spot in toxic releases owes largely to one particular operation on the extreme northwestern coast of the state, more than five hundred miles from the nearest oil installation. Close to the shores of the Chuckchi Sea in this desolate wilderness is the world's largest zinc mine. Named after the pet of a local aviator, the Red Dog mine is a monumental achievement of engineering, but also a stark reminder of the impacts resource extraction can have. An estimated 500 million pounds of "waste rock" is dumped each year into permitted facilities and stored in stockpiles and mine tailings on site.

Nevertheless, the mine provides precious livelihoods for an extremely remote community and is partially owned by the NANA Regional Corporation, a firm belonging to the local Native Alaskan people. From 2000 through 2006, the mine produced an estimated $6 billion worth of zinc, lead, and silver—about 80 percent of the value of all mine output in Alaska. The mine's landlord, NANA, receives royalties from the mine, and under federal law the Kotzebue-based Native corporation shares its payments as well. Sentiments regarding the mine are widely divided between those who benefit from the employment and reside in the nearest town of Kotzebue, eighty-six miles from the mine, and those in small villages closest to the site. A classic "not in my backyard" dilemma is playing itself out in spite of the arrangements the native corporation reached with the mine's Canadian operating company, Teck Cominco. Here, too, issues of governance can confound perceptions of environmental harm. Although the native corporations are focused on maintaining economic development for the communities, the tribal councils are the governance structures with direct connections to the communities in terms of traditional hierarchies. Experts on both sides are being paraded for stories in newspapers as the mine plans an expansion, which might allow it to continue operations until 2031. In an attempt to trump the debate, a former environmental director for the company wrote a rather forceful environmental indictment of the operation in May 2008. The battle rages on, and is being replicated in mining operations worldwide.

Resistance to resource extraction projects is global and highly organized, with environmental groups including Earthworks in the United States, Miningwatch in Canada, and the Mineral Policy Center in Australia devoting their full attention to the issue of the impact of extractive industries on the environment. Several of these movements have also joined forces with anti-globalization activists in critiquing the work of international financial institutions in terms of environmental accountability.[2] In many cases, they have linked with indigenous rights groups that may have misgivings about the pace and manner of development as well. Although the spectrum of opinions among indigenous communities may be highly variable, alliances like these can be opportunistic and not always mutually advantageous, but they are now well established through such leading activists as Winona LaDuke.[3] High-profile cases of mining pollution violating both environmental laws and indigenous rights have given these movements impetus.

Consider, for example, the impact of uranium mining on the Navajo communities in Arizona and New Mexico. Motivated largely by the health impacts on the native reservations, Congress had to enact a separate compensation law for all uranium mines in 1990. As of July 2008, the federal government had paid more than $1.3 billion to persons with lung cancers and other serious diseases as a result of exposure to radiation, with almost half a billion going to miners.[4] In South Africa, up to one hundred thousand apartheid-era mine workers have suffered severe pulmonary ailments from breathing silica dust during gold mining and sued various mining companies for compensation as well.

Apart from the ongoing impacts of extraction, there is also the much more intangible question of risk related to accidents in extractive projects, which is hard for many communities to comprehend. For example, even after much of the uranium mining on Navajo lands had been stopped, a tailings dam failure in 1979 released eleven hundred tons of radioactive mill wastes and ninety million gallons of contaminated liquid, which caused untold health impacts that are still being recorded in nearby communities. With rising prices for uranium, the prospect of reopening the uranium mines in Navajo territory was entertained again, as it might bring an influx of new revenues to the community. The risk has so far been deemed unacceptable by the Navajo people, and their president, Joe Shirley, stated unequivocally in an interview on National Public Radio: "We just don't want uranium mining again—period."[5]

The fundamental premise remains the same: mineral extraction will undoubtedly have environmental impacts and society will inevitably have to grapple with what represents a tolerable risk for the benefit at stake. Would the impacts be reversible? If so, over what timescales and to what extent? What alternative livelihoods would be affected by the operation in the present and into the future? These are all questions that modern society is at least beginning to ask more cogently. This is a departure from the early days of large-scale industrial mining, when environmental factors were hardly on the horizon.

The Scars of Success

Adolf Luderitz, a merchant who played an instrumental role in colonizing present-day Namibia for the Germans in the late nineteenth century, was intrigued with the prospects of minerals in this desolate land of towering dunes and vast expanses of steppe. During his travels from the Skeleton Coast to the interior he commented in one of his journals: "I should be pleased if it turned out that the entire [colony] is a colossal mineral deposit, which once it is mined will leave the area one gaping hole!"[6] Such was the insatiable desire of the colonial enterprise that environmental factors were not even remotely considered by merchants like Luderitz, who took pride in the prospect of such monumental scarring of the landscape. Gaping holes were indeed a literal impact of the mines in Africa that alarmed early travelers to the region from Europe. The Victorian novelist Anthony Trollope recounted his first sight of the great mine at Kimberley in South Africa in his famous work *The Way We Live Now* as follows: "This is the Kimberley Mine. You immediately feel that it is the largest and most complete hole ever made by human agency."[7]

Such sentiment was also expressed by chroniclers of resource extraction efforts in the Orient. In a petition to the ruling shogun, a Japanese finance minister wrote in 1706: "A thousand years ago, gold, silver, and copper were unknown in Japan, yet there was no lack of what men needed. The earth was fertile and produced the best sort of wealth. Gangin was the first prince who caused the mines to be worked diligently and during his reign so great a quantity of gold and silver were extracted from them as no one could have any conception of, and since these metals resemble the bones of the human body inasmuch as what is once extracted from the earth is not reproduced."[8]

The environmental impact of mining was recognized even by some of its most vocal proponents. The German metallurgist Georgius Agricola said in his magnum opus *De Re Metallica* (published in 1556) that "the strongest argument of the detractors is that fields are devastated by mining operations."[9] Disregard for the landscape was not confined by any means to the Old World. Some of the most visible images of the environmental impact of mining come from the Americas. By one estimate, during the California gold rush more than 15 million pounds of mercury used to amalgamate gold were deposited into the watersheds of the Sierra Nevada. Hydraulic mining, which involved the use of high-pressure water hoses to dislodge gold-bearing soil, washed away an estimated 1.5 billion cubic yards of rock and earth, leading one observer to comment that "nature here reminds one of a princess fallen into the hands of robbers who cut off her fingers for the jewels she wears."[10] Fortunately, there was growing concern about the devastation being caused even in those days of relative lawlessness. The farmers who were often victimized by hydraulic mining sued the miners polluting the rivers, raising riverbeds, and making their farms vulnerable to flooding, winning a landmark victory in a case known as *Edwards Woodruff v. North Bloomfield Gravel Mining Company,* decided in the United States District Court in San Francisco by Judge Lorenzo Sawyer in 1884. The decision declared that hydraulic mining was "a public and private nuisance" and prohibited its operation in areas tributary to navigable streams and rivers. In 1893, however, the United States Congress passed the Camminetti Act, which allowed such mining if sediment detention structures were constructed.

As early as 1893 there was a statute on the books in Wisconsin that prohibited the pollution of streams with "saw dust, lime, or other deleterious substances."[11] One of the earliest pollution cases to reach the highest courts in the United States involved mining. Two residents of Scranton, Pennsylvania, J. Gardiner Sanderson and Eliza Sanderson, sued the Pennsylvania Coal Company in 1878 for polluting a brook that flowed through their property. Seven years later, when the state supreme court rendered its verdict, the justices favored the mining company on the principle of "greater good." Environmental degradation was then deemed by the justices "a trifling inconvenience to particular persons . . . which must sometimes give way to the necessities of a great community." Almost a hundred years later, the National Academy of Sciences conducted a study of mining lands, with particular reference to the strip-mine sites, and concluded that it might be best to declare them "National Sacrifice Areas."

When industrial mining overtook small-scale operations at the turn of the twentieth century, the impacts were amplified severalfold. The world's largest single human excavation is the Bingham Canyon mine in Utah, which has been in operation since 1906 and now has a pit a mile deep and two and a half miles across, covering an area of nearly two thousand acres. The sight of the canyon inspires anxiety and admiration at the same time, for it is a clear symbol of the power of humankind in altering the landscape. In comparison, the Grand Canyon is about a mile deep and up to eighteen miles wide, but it took the Colorado River more than 6 million years to achieve that scale with the help of faulting. The cavity at Bingham Canyon is almost as deep, but it was carved in less than a century. Surface mining impacts are perhaps best measured at an aggregate level, both literally and figuratively, by looking at the waste these operations generate. Amounts of waste vary dramatically according to the type of ore mined, so environmental concerns are quite different, depending on the minerals being sought (fig. 11). In addition, changes in hydrology, erosion, land subsidence, and dust are significant concerns that need to be accounted for in any cost-benefit analysis of extraction operations.

Copper
Gross production = 6,780
Net production = 15

Coal
Gross production = 26,668
Net production = 5,476

Phosphates
Gross production = 567
Net production = 142

Petroleum
Gross production = 3,883
Net production = 3,821

Bauxite
Gross production = 533
Net production = 178

Building stone and aggregates
Gross production = 16,950
Net production = 12,463

Iron
Gross production = 431
Net production = 82.5

Gold
Gross production = 2,332
Net production = 0.002

Zinc
Gross production = 319
Net production = 99.8

Brown coal and lignite
Gross production = 9,900
Net production = 1,000

Nickel
Gross production = 883
Net production = 1.6

Clay
Gross production = 220
Net production = 146

Mineral produced | Waste

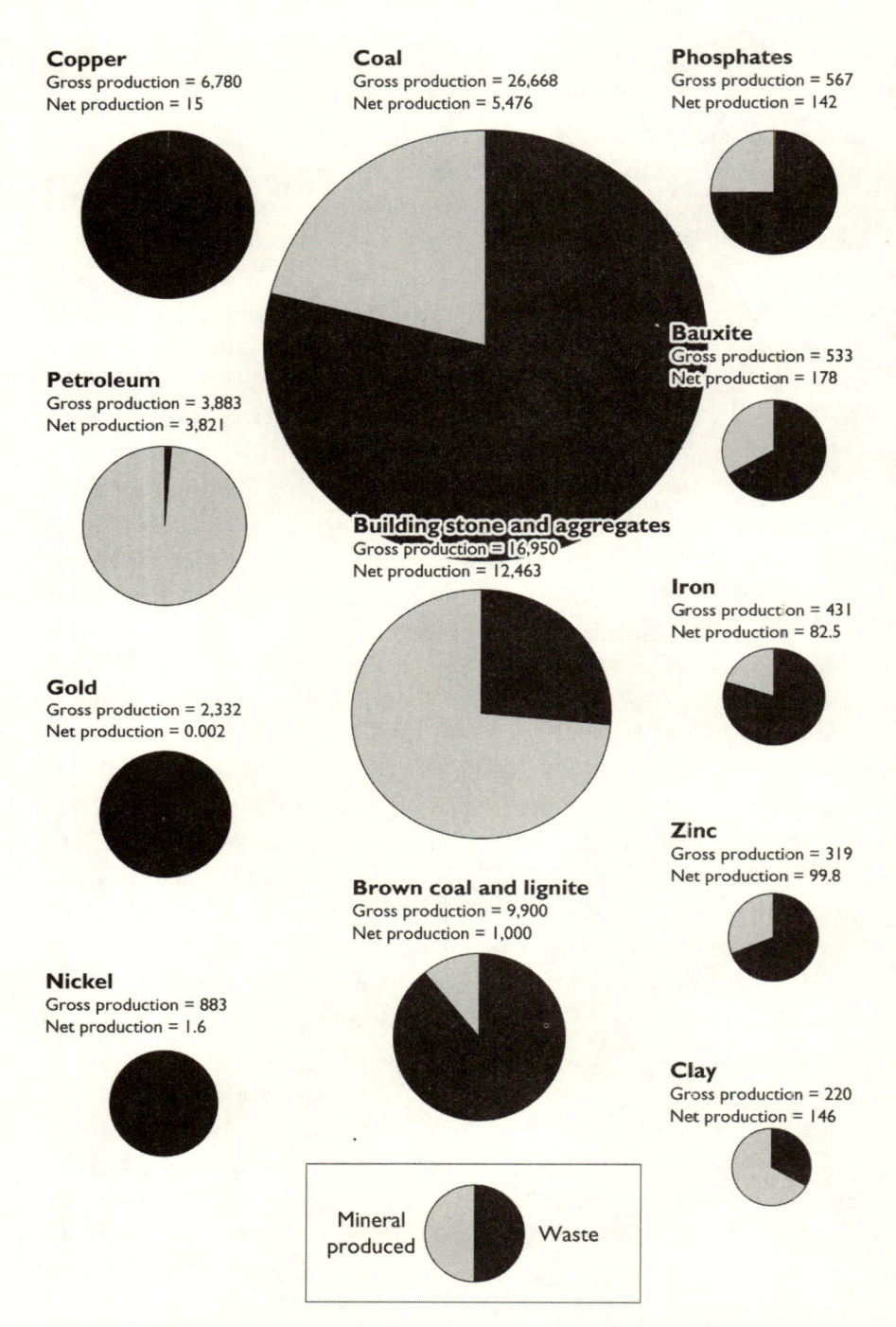

FIGURE 11. World mineral production and waste generation. Figures are in millions of tons. (Data compiled by Nigel Lawson, Department of Geography, University of Manchester, England, 2007)

Usually the ore is covered by rock material of no economic value that has to be physically removed to get at the raw material containing the minerals. Underground mining may not make the same physical scar on the landscape as surface mining, but many of the pollution concerns are similar, and the occupational safety hazards may be even greater.

The ore, once extracted, has to be refined in some form or another, either through physical separation or through chemical mechanisms. For coal, aggregates, and diamonds, the refining is usually done through physical processes of washing and gravity separation of impurities or the use of lasers and x-ray technology. For metals, the refining or "beneficiation" may involve either of two key processes: hydrometallurgical methods that involve leaching the metal out with solvents, or pyrometallurgical processes that involve furnaces for smelting the ore. The use of flotation processes to separate some impurities followed by a series chemical reactions with compounds may be necessary in either process. This might be followed by electrolysis techniques, involving acidic solvents, to get a higher degree of purity of the metal.

The wastes that are produced as a result of the refining process are called tailings, which often need to be stored on site in huge impoundments known as tailings dams. The scale of the tailings facilities can often be much larger than the mining excavation itself. For liquids such as oil, the refining is done through various levels of distillation procedures in a refinery that can be physically located thousands of miles away. The impacts of transportation by pipeline, tankers, or surface vehicles must also be considered. In the case of tar sands, there is an additional step required to separate the bitumen from the sand, which involves large quantities of water being used. The waste water from this process can itself be a major environmental hazard, since wildlife can mistake the ponds for lakes. In May 2008, more than five hundred ducks perished near Fort McMurray, Alberta, when they landed on such a wastewater pond. This prompted several organizations in the United States to boycott Albertan oil, including a declaration by mayors of prominent American cities.[12]

Large-scale mining operations that remove entire mountaintops are now a cause for major outrage in Appalachia, where low-grade coal, previously mined underground, is still being harnessed. In this case, even members of long-time mining families are beginning to show grave concerns about the impact of this kind of mining. In 2003, the highly acclaimed Goldman Environment Prize (given to an activist at a global scale) was awarded to Judy Bonds from the community of Coal River in West Virginia. The environmental impact of coal mining in Appalachia has been exacerbated by the lack of awareness in the region. According to sociologists who have studied underdevelopment in Appalachia for decades, the structural problems with wealth distribution led to a society that was often unable to develop its educational capacity. This, in turn, made communities less able to resist environmental harms, just as much as it made them relatively quiescent in the face of economic oppression for decades.[13]

Judy Bonds, however, defied the odds and took on the challenge of campaigning for an end to mountaintop-removal mining. Even though she was from a family that had mined coal for ten generations, the destruction she was seeing compelled her to act against the industry. For her the mining processes had reached a point when it was simply time to stop. All the easy coal was gone, and it became clear to her that getting at what remained was simply not worth the environmental cost.[14] Entire valleys were degraded due to changes in hydrology as huge waste-rock and tailings dams had to be constructed to store what was left over. Her environmental cause is reminiscent of the great labor activist Mary Harris Jones (better known as Mother Jones), who fought for Appalachian miners' rights in the late nineteenth century and was called "the most dangerous woman in America" by Reeze Blizzard, a West Virginia district attorney. Although the rhetoric of such leaders may seem extreme to many, there are times in assessing environmental damage where compromise might mean that there is an irreversible impact on the ecosystem. If long-term livelihoods of communities are threatened in such cases, confrontation through political activism or legal action becomes the only recourse.

In Appalachia, groups like Coal River Mountain Watch, which Judy Bonds helped to found, are taking on the battle in U.S. courts. However, in many other parts of the world domestic litigation might not be empowering enough or adequately enforced to make much difference. International campaigns are then essential to ensure environmental accountability across geographic boundaries. The Ok Tedi copper and gold mine in Papua New Guinea is a notable example of transnational advocacy, in this case focused on a parent company in Australia, eventually forcing changes and leading to a legal settlement for environmental harm worth over $28 million. A few years after the settlement, the company admitted that the environmental damage caused by the mine was unacceptable and would not be allowable under current standards for mining companies. The case was particularly important because the mine was studied for years by anthropologists, and there was considerable debate about the interventionist role that scientists should play when they are witnesses to such egregious environmental and human rights violations. In his notable new book *Reverse Anthropology,* Stuart Kirsch has made this argument in the context of Ok Tedi and other resource extraction efforts in Papua New Guinea, a land that captures the naturalist's imagination with the prospect of "uncontacted" tribal populations and birds of paradise.[15]

Without appropriate safeguards, extractive industries scar not only the earth but also the communities where they are situated. This was the case with another mining project in New Guinea on the island of Bougainville, which has become synonymous with mineral conflicts in the Pacific and has been the subject of documentary films, monographs, and even a teaching case study by Harvard Business School. The complex interactions between the company, the islanders, and the federal government defy clear culpability, but the community demand to stop mining altogether as part of the peace negotiations shows that the impact of the extractive sector was profound at multiple levels for the residents.[16] Perhaps the most potent metaphor for the impact of mining without appropriate social safety nets can be found in the scholarship of anthropologist June Nash, who studied tin mining and the indigenous populations of Bolivia. Her book *We Eat the Mines and the Mines Eat Us* has become a classic in social anthropology, as it captures the starkness of the relationship between the miner and the mine. Following an accident that took the lives of three miners, she recalls witnessing a ritual in the town of Oruro where a llama was sacrificed as an offering to Supay, the hill spirit whom the miners believe controls the wealth of the mines "so that he will continue to reveal the veins of metal to us and so that we can live."[17] The lethal necessity of extraction has always been a recurring theme in mining narratives of the workers themselves. Understanding the stories of the miners and workers who toiled tirelessly in the bowels of the earth at great risks is imperative in appreciating our complex relationships with minerals.

The Miner's Bird

Observing the subterranean labor force that fueled the industrial revolution in England, Friedrich Engels wrote about the "mining proletariat" that "in the whole of the British Empire, there is no occupation in which a man may meet his end in so many diverse ways as this one. The coal mine is the scene of a multitude of the most terrifying calamities, and these come directly from the selfishness of the bourgeoisie."[18] It is highly ironic that although the fathers of communism identified such issues of mining health and safety in their earlier writings, the implementation of communism in much of Eastern Europe and Russia led to despicable working conditions for miners. The larger industrial enterprise completely subsumed much of the positive rhetoric on workers' safety. Capitalism and communism shared the common infamy of having dreadful records on miner safety, particularly in the realm of coal.

The hazards in coal mining, especially underground, are numerous and hard to contain. First, there is the issue of ventilation and the potential for being deprived of oxygen, either by lack of air intake or the generation of heavy gases that prevented oxygen from reaching the lungs. The earliest safety measure was the now proverbial canary in a coal mine, used to detect dangerous levels of methane and carbon monoxide, and a classic example of how human health and safety can be at odds with environmental or animal rights goals. From 1911 until 1986, canaries were employed in coal mines across Britain and are believed to have saved the lives of thousands of miners. At their peak in the 1920s, Britain's mines employed more than 718,000 workers, and these miners were a major political force for much of the twentieth century.[19]

Apart from the hazards of suffocation, miners also had to balance the need for light with the constant fear of fire from igniting a mixture of gases called firedamp in the shafts. Coupled with stored dynamite used for blasting tunnels, the fires could be explosive and the impact could even bring down an entire mine's external infrastructure, as happened in Aachen, Germany, in 1930 (fig. 12). During the early decades of the nineteenth century, miners used candles and oil lamps for illumination, and these were particularly susceptible to ignition. After a fire, miners were also exposed to the phenomenon of "afterdamp," in which a toxic mixture of carbonic acid and

FIGURE 12. Aftermath of a mining accident in Aachen, Germany, 1930 (German Mining Museum, Bochum)

other combustion byproducts would poison them. The invention of safety lamps was a pivotal milestone for occupational safety in mines, as well as for the productivity of coal mining and industrial growth more generally. The Cornish scientist Sir Humphrey Davy is credited with inventing the most effective of these lamps, and he is fondly remembered in coal-mining lore for his refusal to patent the product so that it could be affordable and easily available to miners.

Working in the mineral underworld historically exposed miners to pathogens that most other human populations did not have to worry about. The closed working environment and proximity of miners to each other in the line of operations gave pathogens an ideal opportunity to spread. However, many of the diseases of miners came from inhalation of fibrous particles from drilling that was often accentuated by working under conditions of high underground pressure and heat. Of particular note in this regard was a disease called phthisis, commonly known by the epithet of miner's disease. The condition frequently developed into tuberculosis or pneumonia as well.

Children were often employed in coal mining across Europe; because of their short stature they could creep into smaller tunnels. Spending extended periods of time in the dark left the children deprived of sunlight, which helps to generate vitamin D in the body. A deficiency of vitamin D led to a loss in bone integrity and rickets. Apart from the miners themselves, an industrial landscape that depended on coal for development in the early twentieth century was generally deprived of functional sunlight. The hazy vistas of London and Manchester that inspired Impressionist painters like Monet were also the cause of many afflictions, including rickets. A British government report in 1918 found that no less than half the population of Britain's industrial areas suffered from rickets, and referred to the disease as "probably the most potent factor interfering with the efficiency of the race."[20] In addition to child labor, women also played a significant role in coal mining across the world and often received less compensation than male miners. There continues to be a strong concern about gender discrimination in mining communities in developing countries.[21]

The plight of miners was considered by various organizations and philanthropists as early as the 1830s. Often comparisons of coal mining in England were made with colonial oppression and slavery abroad. In one such instance, a wealthy manager of an agricultural estate in Leeds, Richard Oastler, wrote: "The pious and able champions of Negro liberty and colonial rights before they have traveled so far should have sojourned to our immediate neighborhoods. Thousands of our fellow creatures and fellow subjects, both male and female, the inhabitants of Yorkshire town, are at this moment existing in a state of slavery." The outcry about the occupational conditions of miners in Britain led to social science studies (perhaps the earliest detailed environmental health questionnaires were designed in such studies) to ascertain public attitudes toward coal mining as well as the views of the miners themselves. A royal commission established to investigate the plight of the collieries of England published a report of almost two thousand pages in 1842.[22]

Since the advent of industrial mining, temperature and pressure have also been recurring concerns for the health and safety of miners. Some South African mines are as deep as 11,500 feet, and with the rising price of gold they may go deeper still. At such depths the possibility of being marooned because of a technical problem is a recurring fear. In 2007, a power failure at Harmony Gold's Elandsrand gold mine in South Africa trapped 3,200 miners underground for more than twenty-four hours. In other cases mines are being developed in extreme environments, such as arctic regions or at very high altitudes. In one study of a Chilean mine at 15,000 feet above sea level, researchers found that daily altitude variations endured by the miners (whose dormitories were 4,800 feet lower than the mine) were resulting in chronic intermittent hypoxia.[23]

Communicable diseases are also more easily spread in mining environments because of the close proximity of workers. Historically, a major problem was sexually transmitted diseases, since many male miners lived in isolation from their families for weeks or months and indulged in risky sexual behavior. Such incidences have been greatly reduced in recent years as many operations allow for family housing or have a "fly-in fly-out" schedule of a week to ten days per shift. However, there are still strong clusters of HIV-AIDS prevalence in mining communities in Africa. Ashanti Goldfields, Ghana's main gold-mining company, decided to put condoms in every weekly pay packet of miners and claims to have reduced the HIV-infection rate by 75 percent.[24] In South Africa, introducing miners to voluntary community organizations was an effective means of reducing HIV prevalence.[25]

In the case of small-scale gem and gold mining, the abandoned pits can be a breeding ground for mosquitoes and the spread of malaria, dengue fever, and other pathogens.[26] Some of the persistent heavy-metal pollutants can have health impacts long after mining stops. One study in Jordan found that pollutants from ancient Roman copper mines in the region persist today in local acacia plants.[27] Interestingly enough, even in those days, there were regulations being propounded for health and safety at mines, though with minimal environmental awareness. Two bronze tablets found at slag dumps from Roman mines in Portugal in the late nineteenth century suggest that there was a Roman mining law in the first century. It was forbidden to extract ore from the pillars in mines because of safety concerns. However, there are also indications that during periods of conflict, mining conditions deteriorated considerably, and being sent to a mine became tantamount to capital punishment (*damnation ad metalla*).[28]

Apart from the health impact, seismologists have expressed concerns that some very deep mines and drilling activities can trigger seismic activity. Gas drilling in Central Java, Indonesia, is now believed to have triggered the eruption of a "mud volcano" in 2006 that displaced more than fifty thousand people. While initially the eruption was thought to have been the result of an earthquake, more detailed peer-reviewed research into the incident revealed that it had been caused by the sudden influx of gas after drilling into the mud chamber of the volcano.[29] Indonesia's government has subsequently ordered the company to pay 3.8 trillion rupiah ($406 million) in compensation to the victims and to cover the damages.

The litany of occupational hazards in mining is long and convoluted, but we have certainly traversed many hurdles in moving toward a safer industry. Yet accidents in the realm of the underworld, even when they are infrequent, often feel more sinister and compelling and remind us of the sacrifices of those who work in this sector.

Three tragic mining accidents in the Appalachians in 2006 struck a sympathetic chord with Americans. Although President Bush expressed concern about the nation's "addiction to oil" in his State of the Union address that year, one of his proposed solutions was to mine more coal. Regrettably, environmental and social policy tends to be reactive rather than proactive, and there has now been a flurry of efforts toward a legislative response to mining accidents that might give the industry a clean bill of health. Soon after the Sago Mine tragedy, which killed twelve miners, the Senate hurriedly held a hearing on mine safety on January 23, 2006, but the government seemed intent on downplaying the matter. David Dye, the acting director of the Mine Safety and Health Administration (MSHA) at the time, walked out halfway through the two-hour Senate hearing. When asked by Republican senator Arlen Specter, chairman of the Senate Appropriations Committee, to stay for the full hearing, Dye responded that the hearing had diverted him and other MSHA officials from more pressing matters, including a mine fire in Colorado, which incidentally had been burning since November of the previous year. A subsequent hearing on mine safety was held on March 1, 2006, but was adjourned by the chairman, Representative Charles Norwood (R-Ga.), when more pointed questions about the industry began to be asked. This led to an angry exchange between participants on the hearing floor.[30]

Such impatient conduct is emblematic of policy ambivalence when it comes to minerals, the environment, and occupational health, because the occupational accident data are often manipulated by either side to seek political advantage. A study by Ken Ward of the *Charleston Gazette* published a detailed review that clarifies the situation. Between 1996 and 2005 there have been 297 fatal coal-mining accidents that killed a total of 320 workers. The Mining Safety and Health Administration has fined coal operators more than $14 million, but most of these fines have been reduced to trivial amounts in the courts without clear legal guidance. For example, when 13 miners were killed in the Jim Walter No. 5 explosion in Alabama, MSHA fined the company $435,000. A judge subsequently reduced the fine to $3,000. Such examples show that there are still some problems with compliance assurance. However, the problems in this regard in countries like the United States are still orders of magnitude less than in many other parts of the world. Considering global demographics, it is essential that environmental health and occupational safety be studied and considered in comparison with areas of greatest growth in extraction as well as consumption patterns.

The Dragon's Excrement

Disenchanted by the mess that oil had made in his country, Venezuela's former minister of mines Juan Pablo Pérez Alfonzo is believed to have referred to the precious liquid as "the devil's excrement."[31] Perhaps carbon dioxide might be more aptly given this epithet, because the infamous greenhouse gas is emitted as the end result of oil being digested by our industrial metabolism. Regardless of its authenticity and accuracy, the metaphor is compelling and perhaps nicely transferable to the largest consumer of many kinds of minerals in the world.

The Chinese have always had a rather ambivalent concept of the devil, and even a mythical creature that conjured fire and brimstone in Europe was considered more benevolent in China. The Chinese thought of dragons as redeeming forces and omens of progress and power. The dragon symbolized China as a nation since the advent of the Qing dynasty, when flags with dragon imagery were first used. It is thus quite fitting that contemporary narratives of China's progress frequently use the imagery of a "racing dragon" with an alarming rate of economic growth. Yet the ravenous appetite for resources of the great nation that bears this potent symbol must inevitably have some consequences.[32]

Domestic minerals fueled much of China's early industrialization and were a major component of Mao Zedong's early plans for development of the country. Judith Shapiro, in her volume *Mao's War on Nature,* described how the longtime communist leader had even been willing to donate royalties from his own writing to ensure that the mineral resources of Panzhihua, located two hundred miles north of Yunnan's capital, Kunming, were developed to the fullest extent.[33]

Iron ore has always been the backbone of infrastructure development because of its importance in steel manufacturing. China is the world's leading producer of pig iron and crude steel (34 percent of the global total), with an output greater than the combined production of Japan, the United States, and Russia, which ranked second, third, and fourth, respectively.[34] Steel production is highly energy intensive, accounting for over 66 percent of China's energy demand and resulting in the need for even greater mining of coal.

The impact of the iron-ore bonanza on China's environment can at times have unexpected manifestations, such as the rise of "magnet fishing" in the streams of Anhui province. Community members are flocking to streams with magnets extended on fishing poles to find lumps of iron ore washed down from waste rock at nearby mines, hoping that the rising price of iron will allow them to sell the lumps and earn more than fishing![35]

Often consumption data for China are presented in terms of iron and steel to highlight infrastructure projects and their ecological footprint. However, production information can be just as revealing about environmental impact. From 2002 to 2007, China's share of global iron-ore production has more than doubled, although it is now expected to slow down as stocks are depleted and the government attempts to temper the impact of such rapid growth on the environment and communities. Additionally, the iron content of the ores on average has been declining. Large-scale mines account for about 20 percent of total iron-ore production, whereas medium- to small-scale mines produce the bulk of the ore.

Global attention on Chinese mining has been largely focused on occupational hazards in the coal-mining industry, which has an unenviable safety record. Indeed, the health and safety performance for Chinese coal mining is orders of magnitude worse than global standards. For example, a study by the U.S. National Energy Technology Laboratory in 2006 concluded that for every one fatality per unit of coal produced in the United States, the comparable rate in China is between 200 and 250.[36] Changes in the mining laws, a projected investment of $60 billion from 2007 to 2011 on safety improvement mechanisms, and the closure of around 23,000 small coal mines with poor safety records are expected to make substantive improvements in this arena.[37] The United Nations Development Programme has started a $14.4 million project for improved coal-mining safety education and training in the five Chinese provinces of Anhui, Guizhou, Henan, Liaoning, and Shanxi.[38] According to official data released in 2008, the accident rate has already gone down by 20 percent through these measures, though unions have often disputed the accuracy of accident reporting.[39]

Although progress is apparent on the health and safety front, the environmental impact of mining still remains unresolved. First, data on pollution from mining are hard to obtain because impact categories are highly diffuse and dependent on the kind of mineral being extracted. Second, most of the data on pollution are gathered by researchers in open access areas rather than at point sources. In such measurements, there are often competing sources of pollution, which makes ambient measurements hard to trace to a particular source. Most of the studies in this regard have been done in the southeastern province of Yunnan, which has a large mining footprint and is also a biodiversity hot spot, and hence of considerable research interest.

Although the agricultural contribution to Yunnan's GDP has been declining, there has still been an increase in the overall land acreage being used for agriculture, which has led to serious concerns about the overall pollution footprint. A study of agricultural expansion in Yunnan estimated that from 1960 to 2003 the push toward increased crop yield led to a ninety-nine-fold increase in fertilizer usage and a doubling of pesticide usage. Interactions between these agricultural chemicals and pollutants from mines in waterways and lakes make impact categorization more difficult.[40] Careful monitoring of effluent from pollution sources and waterways will be essential in years to come to identify the sources of major impact and to direct enforcement action or policy reform accordingly.

One particular area of concern with regard to environmental pollution is the growth of gold mining on both large and small scales. During the height of the communist era, gold mining was banned as a diversionary extravagance, but the extraction of gold can be traced back in Chinese history to the Song dynasty in the eleventh century. Following the advent of economic liberalization policies in the last two decades, the value of gold has been rediscovered by Chinese consumers. Since 2007, China has become the world's largest gold producer (surpassing South Africa, which had held the top spot since 1905), to meet its demand for the prized metal, which jumped 23 percent, making it the world's second-largest consumer after India. Mercury usage for the amalgamation process to extract gold is widespread in China at all scales and is a major concern for bioaccumulation in freshwater fish, which are a primary source of protein for many communities. Although mercury is generally well recovered in large-scale mining processes that use amalgamation, which emit approximately 0.79 ounces of mercury for every ounce of gold they produce, far less of the toxic metal is recovered in small-scale mining, where emissions are roughly 15 ounces of mercury for each ounce of gold.[41]

Mines that do not use mercury rely on cyanide for processing the gold, and this usage has its own share of environmental concerns. Cyanide decomposes fairly fast in sunlight, but it is still a potent toxin and can cause major damage without appropriate management practices. There have been several accidents in China involving cyanide spillage in recent years. In September 2007, a house built on top of a cyanide pool in a gold-mining area of Henan province collapsed, and nine people died after falling into a cyanide solution six feet deep. The tale of this tragedy is particularly poignant, as the villagers had initially gathered in the house to discuss funeral arrangements for a young man who had been found dead in Yangping village (near the city of Lingpao). Apparently, the youngster had quarreled with his parents the previous day and run away to spend the night alone in an abandoned shack, where he too was most likely poisoned by cyanide fumes.[42] This was by no means an isolated incident. A hydrogen cyanide gas leak from a gold-mining plant in Beijing's suburban district of Huairou killed three people and left another fifteen hospitalized in 2004. Three years earlier, eleven tons of liquid sodium cyanide leaked into a tributary of the Luohe river in Henan province.[43]

With thousands of mines all over the country in these boom years, China will also need to consider a long-term strategy for dealing with post-closure environmental impacts. Communities that have relied on mining for generations need to be made aware of when the deposits are going to be potentially exhausted years in advance, so they can plan for effective alternatives and gradually transition their livelihoods.[44]

Without such planning, companies can walk away from a legacy of pollution and a complete vacuum of economic alternatives with impunity. Consider the example of mercury mining in China, which officially ceased in 2001 after six hundred years of operating all over the country, particularly in Guizhou, Shanxi, Henan, and Sichuan. More than 100 tons of calcines and other waste rocks have been produced as a result of the mercury mining in Wanshan, which was the mercury capital of the country. As a result of the mining in the surrounding areas, soil concentrations range from 16 to 232 times the maximum amount of mercury allowed for soils in China. Concentrations in adjoining waterways were also found to be highly elevated more than six years after the mining stopped. Because of the biological persistence of mercury, concentrations of methyl mercury in grains grown in this region are also extremely high.[45] Yet there is no mechanism to hold companies to account for this contamination, and the government will have to bear the burden of cleanup, at a time when it is eager to start newer projects.

There is certainly far more mining that is likely to occur in China this century, and it is clear that past mining practices cannot be sustained in the future. The government has identified 60.7 billion tons of iron-ore reserves, and is expected to identify more than 100 billion tons with better exploration techniques. The Ministry of Land and Resources is now using international models to explore new deposits, which could increase estimates of China's identified reserves of copper ore by 20 million tons and bauxite by 200 million tons by the end of 2010.[46] All of the new projects that will commence in coming years for these reserves will need to be planned with greater care if the country is to avoid an ecological crisis of immense proportions.

The first step to improve environmental performance in the Chinese mining industry will entail an exertion of regulatory authority over all mining areas. Between 2005 and 2008, the Ministry of Land and Resources has identified more than sixty-five thousand unlicensed mines, forty-five hundred unauthorized excavations, about a thousand unauthorized prospecting sites, and thirteen hundred illegal transfer issues of mining rights, and it has shut down more than eight thousand illegal mines.[47] The illegal mining activity is also taking a toll on cultural sites. In February 2008, the government arrested twenty-six people suspected of illegally mining and damaging a protected six-thousand-year-old Neolithic site in the northeastern province of Liaoning (short-listed as a UNESCO World Heritage site). The government was also forced to ban mining near Buddhist mountains and shrines after severe protests from Tibetan monks in Shanxi and Sichuan in early 2008.[48]

A coherent policy will also be needed to institutionalize the small-scale mining sector in China, which employs an estimated 6 million people and accounts for large shares of mining in many sectors. An estimate by a Canadian researcher in 2005 concluded that small-scale mining accounts for more than 75 percent of China's bauxite production, 65 percent of manganese, 51 percent of phosphate, and 43 percent of coal.[49] With the help of the British government and the World Bank, there has been some movement to mainstream this sector through the establishment of a Chinese branch of an international program called Communities and Small-scale Mining (CASM). Led by researchers at the Chinese Academy of Sciences, this program aims to focus on this sector to increase its potential for providing livelihoods as well as improving its environmental and social performance.[50]

The primary resource sector has played a pivotal role in propelling China to Olympian heights of development by most measures. The ubiquity of goods made in China is a hallmark of the cultural spirit of this great nation, and how it has become the fulcrum of modern materialism. The dragons that embellish Chinese restaurants in even the most remote communities across the world are emblematic of the cultural power of China. Yet consumption of all resources, even as basic as food, inevitably passes through the system and has some impact on the environment. As the world's most populous country, China will undoubtedly have to play a crucial role in improving the ecological performance of its primary resource sector to benefit us all.

The Price of Depletion

The statistician Harold Hotelling is fondly remembered by many mining economists for grappling with that most imponderable of questions for resource extractors, which also troubles environmentalists: What should we extract now versus in the future, knowing that the resource is finite? Through a detailed mathematical analysis, he showed as early as 1931 how we could in fact reach an "optimal rate of depletion," given various considerations of extraction costs and the price as determined by the number of firms in the market. The key determining factor in his analysis was the discount rate—a measure of how much we value the future compared with the present. The concept was clearly a problem from an environmental perspective, since it deals with the tricky issue of intergenerational equity. Assigning a discount rate means assuming that parents should be able to estimate the worth of a resource for their children, but even more consequentially for future generations in perpetuity. Yet the reason for having a discount rate is that there is indeed uncertainty, for better or worse, in all planning endeavors. The issue of technological progress remains a major unknown, and ideally the rate of depletion of a resource would somehow keep up with the time lapse in technological development for alternatives.

Part of the challenge for economists in championing market mechanisms to account for environmental harms of resource extraction has been the inability of markets to account for complex patterns and feedbacks. This is perhaps a reflection of the inadequacies of human behavioral responses to certain ecological cues. The renowned development economist Sir Partha Dasgupta describes this imponderable as follows: "Markets are able to function well only if the processes governing the transformation of goods and services into further goods and services are linear. However, when someone talks of 'ecosystem stress' or 'ecological thresholds' as ecologists frequently do, they mean states of affair in systems governed by nonlinear processes. For example, market prices may be unable to signal the impending collapse of a local resource base."[51]

This was clearly true for the inhabitants of Nauru, a tiny Pacific island nation that for decades was home to one of the most prosperous phosphate mining industries in the world. With fewer than ten thousand people,

Nauru is the least populous member of the United Nations, but its phosphate revenues once made it the richest in the region. A lack of planning by the island's German and Australian colonizers, and subsequently a continued total reliance on one commodity, plunged the island's economy into chaos. The unemployment rate is over 90 percent, and the land has been so severely degraded by mining that alternative livelihoods are next to impossible (fig. 13). So desperate was the country for employment and income that it agreed to host an illegal immigrant detention center for the Australian government, which was operational until 2008. The Australian government agreed to an out-of-court settlement to rehabilitate some of the environmental damage from the phosphate mining, and the long-term future of the island remains dependent on this ecological restoration process.[52]

Economists have successfully branded themselves as scientists with mathematical exactitude who can artfully negotiate the vagaries of human consumption patterns through pricing mechanisms. Yet the life support systems that sustain the planet have eluded their grasp, often being relegated to the

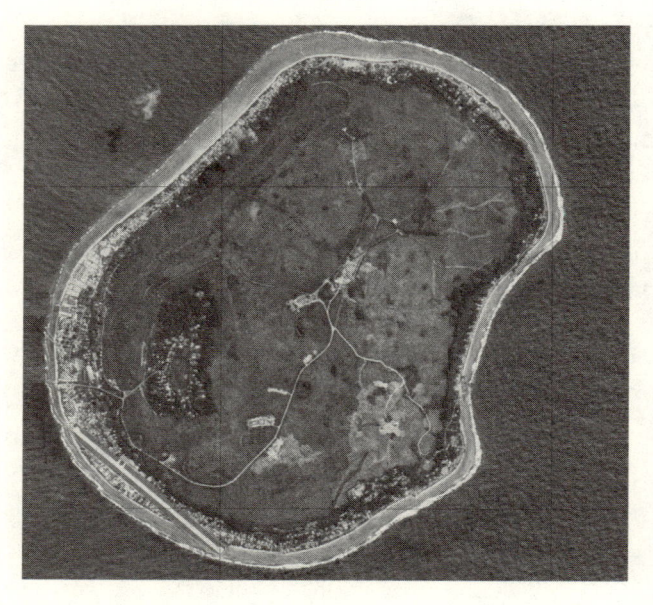

FIGURE 13. The island nation of Nauru in the South Pacific. The central part of the island (shown in white) depicts the extent of the mining that has now left the land barren. (Digital Globe–Google Earth, published in *Alternatives* 35, no. 1 [2009]: 34)

residual category of externalities. Until recently, if you asked a conventional economist about "pricing nature" you risked being ridiculed as a green hippie or a neo-Marxist. Professional economic associations and journals marginalized any dissent from conventional pricing etiquette, which could only be determined under strict market conditions. The most accommodation economists could grant was trading of emission permits for air pollution, and that too was considered a side show within the halls of dominant economics programs.

Given this strong ostracization, a parallel field of ecological economics had to develop, led by a few rebellious researchers. Most notably, the Romanian-American economist Nicholas Georgescu-Roegen, who had been a protégé of Joseph Schumpeter at Harvard, dared to embrace other physical sciences, such as physics and biology, in his analysis. His seminal book *The Entropy Law and the Economic Process* (1971) was the first treatise to consider physical constraints on capitalism described by his mentor, Schumpeter, as "a process of creative destruction." However, at a time when even a nuanced challenge to capitalism was considered the same as communist innuendo, most of the core ideas of such works were dismissed along with more qualitative and normative writings, such as E. F. Schumacher's *Small Is Beautiful* (1973).

Interestingly enough, a challenge to consumer sovereignty had been presented several years earlier by the far more influential economist John Kenneth Galbraith in his work *The Affluent Society* (1958). However, his narrative was considered more in the realm of political science than economics and did not have the ecological cadence that was to follow as the field of ecological economics developed. Disciplinary reductionism has been pervasive in economics far more than any of the other social sciences, primarily because the dominant use of statistical methods made the field esoteric to the public. Meanwhile, the ecological economists matured to build bridges with other disciplines, so much that it wasn't necessary to have a Ph.D. in economics to be labeled an ecological economist. They argued that economics was essentially a study of human responses to scarcity of resources, and hence required an eclectic assemblage of tools and fields of inquiry.

Ironically, ecological economists also began to develop more rigorous mathematical models and followed some of the same methodological paths as their conventional cousins. However, neither side communicated much with the other, and both built ideological walls between the fields. The ecological economists developed their own professional society and journals and were less reliant on the peer review filters of the conventional economists. In the past decade, there seems to be a promising shift across the tectonic plates of economic thought that just might close the fault lines. Ecological economists have moved closer to pricing strategies that have been the pulse of conventional economic analysis. To traverse this journey, they had to encounter criticism from existential environmental philosophers such as Mark Sagoff, who has chided them for selling out and commodifying nature.[53] However, the ecological economists have admirably gone beyond utopianism and moved toward a pragmatic approach to addressing environmental concerns. While they might not have the prices right at this stage, at least they are trying to delineate monetary indicators in tangible terms rather than using the polemics of priceless value.

The diminutive Central American country of Costa Rica is often considered an outlier because of unusual policy decisions, such as not having an active army. However, the challenges and successes of policy reform encountered there are just as real as they would be in other parts of the world. Costa Rica hosted a major international conference on payments for ecosystem services in 2007, and the country's first environment minister shared how difficult it had been for him to initially convince the finance ministry about a system of allocating revenues from targeted taxes to communities that were conserving ecosystems. Once the program was launched, however, there were clear win-win opportunities for the country that became evident from the improvement in development indicators for some of the most remote communities where ecosystems were being conserved. At the same event the former prime minister of Suriname shared how the leading newspaper in his country had dismissed initial efforts at conservation but had since come around to realizing the value of ecosystem protection. The idea had even caught on with BHP Billiton, one of the world's largest mining companies, which was working with the government to establish buffer zones for its new mining projects in Suriname to mitigate ecosystem disruption.[54]

For the skeptics who would relegate such efforts to small countries with unique circumstances, let us consider the case of China's Grain for Green program, launched in partnership with the World Wildlife Fund (WWF). The program has allowed farmers who conserve the environment to be paid through a fund created by a tax on polluting industries elsewhere in the country. By 2010, the program aims to conserve 34 million acres of cropland. The conserved land can still be productive for food crops as long as minimal environmental criteria are met.

Another research effort on payments for ecosystem services is emanating from the biological sciences, led by Gretchen Daily at Stanford University, which is targeting investment banks in an attempt to make such efforts more mainstream in accounting processes. The National Academy of Sciences gave its blessing to such research as well by devoting a special issue of the prestigious journal *Proceedings of the National Academy of Sciences* to this topic in July 2008, guest-edited by the Stanford research group. Conservation set-aside programs have been practiced in countries belonging to the Organization for Economic Cooperation and Development as well, such as the U.S. Department of Agriculture's Conservation Reserve Program (authorized under the Farm Bill of 1985). However, the refinement and targeted implementation of such programs in developing countries is particularly important. The basic premise is not idealistic at all—it is simply to ensure that the values of natural services are not taken for granted. While some economists are intent on dismissing any mention of pricing natural services as neo-Marxist or theoretically untenable in the absence of a clear market mechanism, assuming that nature's services should be absorbed at no cost is also illogical. A former vice president at Exxon in Norway, Øystein Dahle, stated the challenge quite succinctly: "Socialism collapsed because it did not allow the market to tell the economic truth. Capitalism may collapse because it does not allow the market to tell the ecological truth."[55]

Yet an attractive feature of capitalism is its flexibility to allow for such factors to be configured appropriately. Pricing mechanisms are powerful means of channeling human behavior when they accurately account for the desired policy response in terms of natural constraints. Getting the prices right in the context of ecosystems is no doubt an enormous challenge, but one that we can increasingly tackle with the help of appropriate computational algorithms, rapid communication infrastructure, and a general awareness about future risks of resource depletion. Having an effective means of valuing natural systems is perhaps the most constructive path to balancing resource extraction and conservation.

PART III

MEASURE FOR MEASURE

Thyself and thy belongings
Are not thine own so proper as to waste
Thyself upon thy virtues, they on thee.
Heaven doth with us as we with torches do,
Not light them for themselves; for if our virtues
Did not go forth of us, 't were all alike
As if we had them not. Spirits are not finely touch'd
But to fine issues, nor Nature never lends
The smallest scruple of her excellence
But, like a thrifty goddess, she determines
Herself the glory of a creditor

—William Shakespeare, *Measure for Measure,* Act I, Scene 1

7

DESTINATION CRADLE: THE QUEST FOR MATERIAL CYCLING

"What a large volume of adventures may be grasped within this little span of life, by him who interests his heart in everything, and who, having eyes to see what time and chance are perpetually holding out to him as he journeyeth on his way, misses nothing he can fairly lay his hands on!"

—Laurence Sterne, *A Sentimental Journey,* 1768

Deep beneath the city of Detroit, as much as twelve hundred feet underground, is a network of tunnels and chambers that constitute one of the largest underground mines in America, with more than fourteen hundred acres and fifty miles of roads. The mineral being extracted is common salt, which has been extolled throughout human history for its many virtues, from culinary imperative to antibacterial preserving agent. The United States is the world's largest producer and consumer of salt, earning more than $1.2 billion in sales revenues annually. Yet only 8 percent of this salt is used for its original edible purposes. More than 51 percent of the salt is used for de-icing roads.[1]

The salt, which is scattered across the many thousands of miles of snow-laden winter roads, remains in the environment permanently, where its impact on ecological systems will be felt in various positive and negative ways. Slowly, it is dissolved in meltwater and carried in solution down storm sewers and into waterways as ionized particles, most of which will be absorbed by organisms. Perhaps when conditions are ripe, some of the salt might precipitate out of solution and form some deposits nearby, but it will hardly have the potential to be mined again for reuse, at least not within human timescales. The same can be said of the salt we eat, for it will be metabolized and go through innumerable chemical reactions in our digestive tracts and sewage treatment plants. The constituent elements, primarily sodium and chlorine, remain intact through this process, but they are convulsed through so many bonding exercises with other molecules that for all practical purposes they are lost in a chaotic soup of ions and particles.

Whenever we extract materials from the earth, we need to consider what the prospect might be for recovering them again after their use if we are to extend the time of our tryst with a particular planetary resource. For common salt, we are fortunate that the oceans can provide an almost inexhaustible supply if we were willing to expend a reasonable amount of energy to harness the mineral. We may be able to synthesize some materials with relatively low energy expenditures, in cases where they are chemically comparable to mined compounds. Asbestos mining has been reduced globally because of the synthesis of fiberglass substitutes from sand, which is ubiquitous and much less harmful to human respiratory tracts. The valuable properties of asbestos as a fire retardant and insulation material have thus been nicely replicated in a product that doesn't need to be mined. Existing asbestos is not as widely dispersed in the environment as salt, and could possibly be recovered.

The same cannot be said of many other materials that, once dissipated through use, are much harder to recover. Consider the use of such metals as columbium (niobium) and tantalum (collectively called coltan), which are generally mined in Central Africa and used for their unique properties in storing an electrical charge in capacitors for cellular phones.[2] Phones have become so commonplace and inexpensive that they are often discarded with impunity. Recovering the mined coltan from the phones is considerably difficult without a set of clear procedures and incentives for retrieving the devices after their productive use has expired. Companies including Virgin Mobile are beginning to provide postage-paid envelopes to subscribers for mailing back their old cell phones, but much more effort is needed to implement these efforts in the largest growth markets in Africa and Asia.

For most products made today, waste materials are at the very core of the industrial materials cycle (fig. 14). Waste is the point of intersection between commodities at different stages of refinement and the environment.

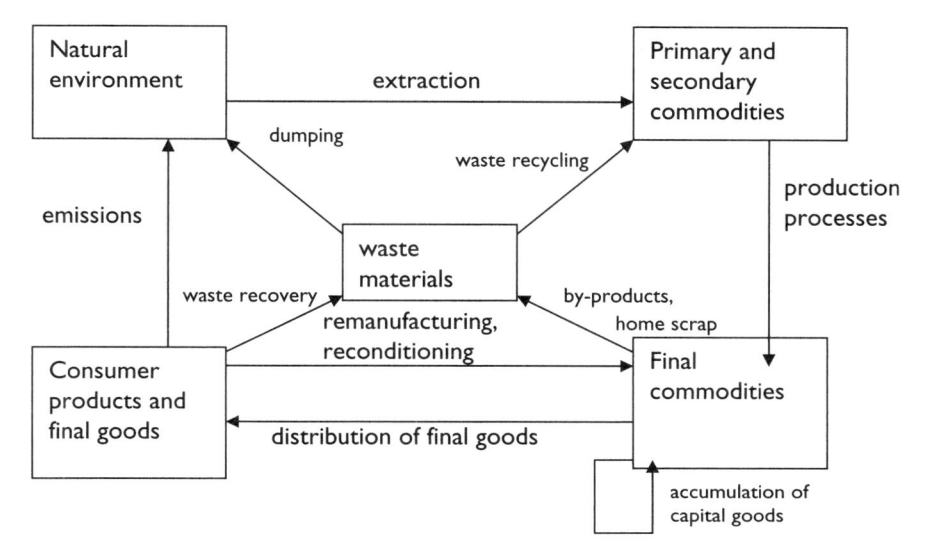

FIGURE 14. Simplified view of industrial material cycles (Jeroen C. J. M. van den Bergh and Marco Janssen, *The Economics of Industrial Ecology* [Cambridge: MIT Press, 2004])

Learning ways of treasuring our wastes is key to meeting one of the most fundamental ecological challenges of our times. Amazingly enough, this challenge is being met with aplomb by some of the most marginalized populations on the planet.

Scavenging Against Chaos

In the northern suburbs of Manila is a twenty-five-acre property called Lupang Pangako, which translates into "promised land" in English. Filipinos are generally very religious people, and one might assume that this is perhaps a shrine to some divinely inspired sanctuary for the destitute. Yet the promises of the hill that this property comprises are all quite human in their origins. Lupang Pangako, or Payatas, as the area is referred to on maps, is home to almost eighty thousand residents who live in the shadow of one of the greatest waste dumps in Asia. What promise did this land invoke for those who named it? Was postcolonial sarcasm at play, or might there be more than meets the eye (and nose) in this fetid agglomeration of junk?

The world's attention was drawn to this site on July 10, 2000, when a landslide in the rubbish dump buried some of the slum settlements surrounding it, and more than three hundred people perished. It was then that the international community began to ask questions about why people were living in such close proximity to a dump, which in most circumstances would invoke a classic repugnance and depreciation of property values. Indeed, many of these residents could have found accommodation farther away but gravitated toward the rubbish for very material interests, enduring the risk of landslides and explosions generated by the methane gas that emanates from the decaying refuse.

The answer was as simple as the rush toward a gold mine. The waste of the elite was the treasure trove of the impoverished in this land of frightful inequality. Residents considered the dump their bank of surprising fortune, where occasionally someone might find an accidentally discarded piece of valuable jewelry or a designer pair of sunglasses that could perhaps be bartered in a pawn shop. The dwellers of Payatas are highly territorial about their stakes around particularly opportune parts of the dump. Refuse from some of the wealthier garden suburbs of Manila that have street names like Harvard Avenue are prime real estate at the dump. The Philippine government has been repeatedly embarrassed by the residents of Payatas, but it has struggled to find a lasting solution. Corrupt rulers have also had a strangely ambivalent relationship with the scavengers. When Imelda Marcos was campaigning to become the governor of metropolitan Manila in the 1970s, she provided electricity to the area, but after the election she became convinced to move toward closure of the site because it gave a negative impression of the country. In 1983, the site was bulldozed, and the residents moved to the township of Bulihan in the province of Cavite. Within three years, however, after the Marcoses were forced out of power and into exile, 90 percent of the residents were back at the dump.[3]

Violent conflict has broken out between residents vying for the most prized trash, and ultimately a formal mediation process had to be launched between the new treasure seekers and the old established slum dwellers. The Catholic church mediated the conflict in the late 1980s and resolved a system of three shifts whereby residents would be allocated time to scavenge for treasure. The evening-shift workers were provided with headlamps amazingly reminiscent of those used by miners.

Depressing as the plight of these waste scavengers may be, the principle behind their purpose is laudatory. Acting in the spirit of primal environmentalism, they are trying to recover usable materials from a high entropy environment, thereby acting to reduce pressures on our resource base. The quest to salvage whatever material may be usable from trash is one that sooner or later all human societies will need to consider if they are to endure growing pressures of population and prosperity. The world's scavenging populations are leading the way in this regard.

Scavenging has been instinctively practiced in human societies since time immemorial. Grave robbing is perhaps the most macabre example of material scavenging, and has been recorded by archaeologists the world over, but intensive forms of material scavenging began with the advent of urbanization. During the industrial revolution, population growth in urban areas increased significantly, and scavenging moved from being criminalized to an acceptable, albeit disparaged, profession. Jean-François Raffaelli's famous painting *Le Chiffonier* (The Rag Picker) immortalized the profession for many in France and across Europe. Scavenging could be highly lucrative, and by one estimate from British engineer Henry Spooner, the value of items recovered by Parisian waste pickers in 1918 totaled around 600,000 pounds sterling (in current dollars, this amounts to around $38 million).[4]

The economic value of scavenging continues to this day. One study among scavengers in Beijing found that their earnings were three times the monthly income of college professors. Estimates at a global level by Martin Medina in his pioneering study of world scavenging practices finds that in the large metropolitan centers of Mexico City, Manila, Buenos Aires, Cairo, and Bogota scavenging brings in a total of $250 million per year. In India, the economic impact of scavenging amounts to around $280 million per year.[5]

Recovering materials from waste is also gaining greater importance with formal institutions. Waste management has been perceived as a rather shady business, often controlled by criminal gangs who capitalize on the essential nature of the service as a public health problem to extort neighborhoods and blackmail governments.[6] This problem was exemplified most acutely by a waste management crisis in Naples, Italy, in 2008, when tons of garbage rotted in the streets because of the municipality's inability to build new landfills and heightened tensions with the Neapolitan Camorra mafia. In the words of one mafioso being interviewed by a magistrate: "For us, rubbish is gold."[7] Unfortunately, the lucrative nature of the trash industry in this context has little to do with recovering economically viable material from waste but rather involves illegally disposing of the material, particularly hazardous waste, which might otherwise be subject to higher disposal fees. Without proper governance, even well-intentioned regulations to encourage proper waste disposal and perhaps recovery can be co-opted by criminal interests that can quash even the potential for positive scavenging.

Configuring the right economic incentives and governance structures for waste management in the most efficient way for producers, consumers, and scavengers (whether formal or informal) is essential. Material usage all over the world has been rising at an alarming rate, and recycled materials still account for a relatively small portion of the overall percentage of mineral usage. Industrial minerals, such as potash and phosphates, which are relatively abundant but also much harder to recycle, account for a major component of mineral usage. In the United States, the use of recycled metals has been increasing since about 1950, and in some cases is even beginning to approach the amount of new metals consumed (fig. 15). The amount of new metal used, however, also continues to increase.

John Seabrook, in his investigations of scrap metal, observes that the metal scrap industry rests on an "inviolable truth: almost all the metal that has ever existed in the world still exists, and always will exist."[8] Yet recovering metals from their various uses remains a challenge in many cases. The irony, quite literally, for environmentalists is that the most abundant of all metals in the earth's crust, aluminum and iron, are also the most easily recycled. The steel industry now brands its product with a registered trademark, calling it "the Enviro-metal," but for those who are most worried about resource depletion, steel was never a priority, because well-established markets for recovery and recycling it already exist. More than 50 percent of all steel produced since World War II has been recycled, with rates in some sectors, such as automobiles and appliances, exceeding 80 percent.[9]

More interesting and perhaps problematic is the recovery of other metals, such as copper, zinc, molybdenum, and uranium. In this context, high

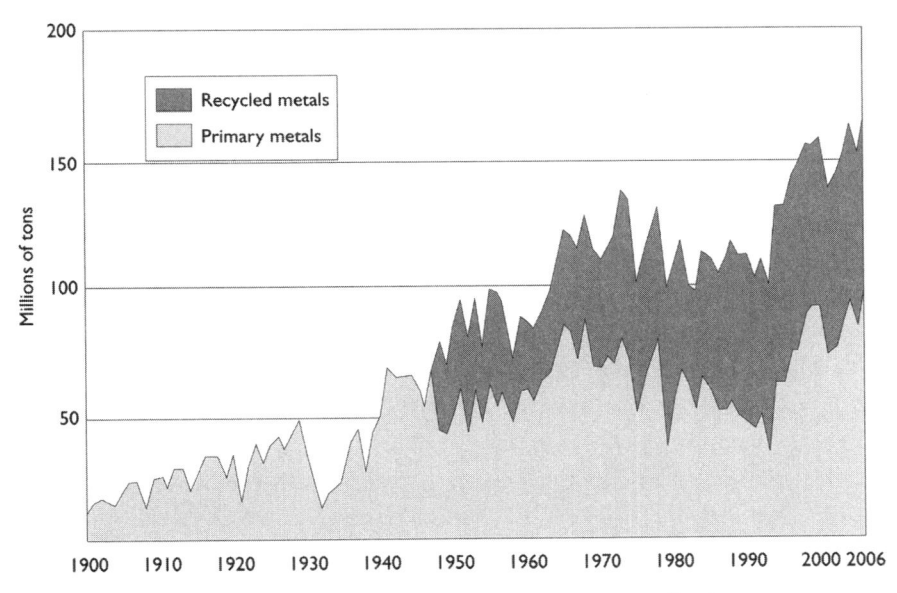

FIGURE 15. Recycled versus primary metals put into use annually, from 1900 through 2006, in the United States (Yale University, Center for Industrial Ecology, Stocks and Flow Project)

levels of corrosion or chemical change make recovery expensive, or the metals have been dispersed into such small, scattered quantities that recovery is impossible. Other metals like gold and platinum are eminently recyclable, but they are often used for durable manufacturing and hence more difficult to recover on purely economic grounds. Commodity prices may change the "want-based" hoarding of such metals as gold and release them for recycling, but there are other institutional challenges, such as the stability of markets or pure sentimentality toward a currency. Consider the lowly copper penny, which was worth more in scrap than its face value when the price of copper exceeded four dollars per pound in 2006. Around the same time, in Bangladesh, old Indian coins with a particularly opportune alloy proportion were being melted down to make razor blades, leading to an acute coin shortage in eastern India. Such price hikes for commodity metals have also led to a surge in scrap theft all over the world reminiscent of dystopic depictions of metal looting in the Australian movie classic *Mad Max*.[10]

In true Taoist fashion, times of crisis can also be a moment of opportunity for the metal recycling business, particularly in China—the world's fastest growing consumer of metal scrap. For example, the destruction of the World Trade Center provided the scrap industry with one hundred thousand tons of high-grade metal, more than half of which was bought by the Shanghai Baosteel Group. An additional ten thousand tons was purchased by scrap dealers in India, which is another major market for scrap.[11] However, much of the scrap in India is used not for domestic infrastructure development but rather to make export-grade steel and products. The country is a destination of choice for the scrap, particularly from battlefields such as Iraq, because of its proximity to supply and demand centers or its abysmally low labor cost. The dangers of a poorly regulated scrap-metal sector in India are particularly acute in the shipyards of the south where huge oil tankers and military vessels from all over the world are dismantled. In China, the major concern is exposure to toxic heavy metals from electronic wastes—more than 70 percent of which are estimated to be sent to China for recovery and disposal. Enforcement of protective equipment for use by the workers, and a clear accountability of how much metal is recovered versus lost is needed to ensure that the scrap sector's environmental credentials are not tarnished.

Much as waste salvaging in all its various forms is a productive endeavor, it still raises the question of why we are producing so much waste to begin with. The amount of material wastage is largely due to a failure to appreciate the full material cycles of products and their inextricable linkage to the world around us. For too long we have assumed that industry is somehow independent of ecological systems. A paradigm shift is now occurring, slowly but surely, which will lead us to consider the role of industrial processes as part of the much larger environmental systems that sustain us all. Valuing waste should be a worthy goal that has precedent in human history at various levels—buildings ready to be demolished at one point in time have been prized as historical treasures in another. TV programs like *Antiques Roadshow* can resuscitate ailing markets in paraphernalia of all sorts. Aluminum, which is presently as commonplace as a soda can, was a prized metal for jewelry in the nineteenth century.[12]

Garbage for some scholars is a treasure trove of data for researching human behavioral changes. Since 1973, William Rathje and his students at the University of Arizona have been going through the trash of Tucson residents to understand patterns of change in behavior and material culture. Eschewing conventional survey techniques, Rathje and his associates believe that "what people have owned—and thrown away—can speak more eloquently, informatively, and truthfully about the lives they lead than they themselves ever may."[13] As an archaeologist Rathje sees trash in the same light as an ancient tomb with hidden artifacts and clues to understanding the mysteries of a bygone civilization.

The wave of planned obsolescence that has struck our society in the past few decades is increasingly evident in what ends up in garbage and data for researchers.[14] Scholars tend to divide obsolescence into three categories: planned failure of materials (disposable products and short-life cheap manufactured goods; functional obsolescence (products outdated by technology); and style obsolescence (products outdated through demands generated by the artistic sector of society and the advertising industry).

Between 1960 and 1990 the intersection of these various forms of planned obsolescence, alongside material efficiency in producing goods, led to a nearly 60 percent increase in waste generation per capita. However, from 1990 to 2006 (and by all indications continuing since then), the per capita waste generation of the United States has more or less plateaued at 4.6 pounds per person per day. Recovery rates for recycling, which increased 8 times between 1960 and 2000, have now leveled off at around 1.1 pounds per day.[15] Do we need more scavenging to reach a higher recycling rates? We probably would not want to have human scavenging in landfills around the United States. Far more efficient mechanisms can be devised for moving beyond the current plateau on recovery for recycling if only we can move toward the emerging new paradigm of industrial ecology.

Industrial Nature

The industrial revolution, which began in Britain in the late eighteenth century, marked a fundamental distancing of human society from the natural system, because it was characterized by a transition from a primarily agricultural economy to one based on manufactured goods. In other words, economic reliance on natural systems as the primary means of production was shifted to mechanized systems of production that were self-perpetuating. Technology gave us the ability to produce finished goods in great numbers and supply them to consumers who were consequently less aware of how the finished product had been manufactured. For example, instead of buying wheat from a farmer or even flour from a mill, you could simply buy large quantities of finished confectionery products and often be quite oblivious of the manufacturing process.

The population began to depend more directly on the firm that supplied the goods rather than the natural system that latently supported the actual enterprise. At the same time, the firm too focused more on the needs of the consumer and took for granted the natural resource base on which it fundamentally depended. The environment was thus short-circuited out of the circle of affluence and prosperity, so long as the resource was available within the desired profit margin. The impact on the natural resource base became an "externality," so long as there was a perceived abundance of the resource.

Around the same time as the boom in coal mining and industrial expansion across Europe, the term "ecology" was conceived by the German naturalist Ernst Heinrich Haeckel in 1869 for the study of the interactions between organisms and their environment. It soon became apparent that organisms and their environment have more than just a biological interaction, so by its very nature ecology was forced to depart from reductionism and offer a more holistic spectrum of inquiry. The concept of an ecosystem, introduced by the British biologist Alfred Tansley in 1935, played a major role in giving structure and coherence to this field. According to ecological historian Frank Golley, "the ecosystem referred to a holistic and integrative ecological concept that combined living organisms and the physical environment into a system."[16] This concept heralded an important realignment in academia from reductionism to holism. Quite surprisingly, the need for holistic approaches to science and policy was also articulated a decade before Tansley's paper by the South African general Jan Smuts in his treatise *Holism and Evolution* (1926).

Ecology, and more specifically the concept of an ecosystem, were thus obliged to consider industrial processes, keeping in line with the holistic, all-encompassing worldview they espoused. The field of cybernetics embraced this holistic worldview from a computational perspective under the direction of the great mathematician Norbert Wiener. Central to the success of computer science has been the significance of feedback loops and networks, which can be traced back to cybernetics.[17] Cybernetics is now an interdisciplinary science dealing with communication and control systems in living organisms, machines, and organizations.

These concepts were further reinforced by the publication in 1971 of Barry Commoner's notable work *The Closing Circle,* which argued for a stronger relationship between modern industrial activity and ecology. The same year the Japanese Ministry of International Trade and Industry (MITI) started a program to promote "principles of industry-ecology," believed by many to be the first formal usage of the concept. This program recognized systems boundaries for industrial activities, and emphasized the development of mechanisms to control human impact on ecosystems and promote ecological equilibrium. This was in sharp contrast to the common modes of engineering discourse that isolated human endeavor and nature as different components with little or no interactions.

The trend toward holistic thought since then has come about largely due to humanity's growing ability to effectively manage large amounts of information. Advances in quantitative analysis and computing are perhaps the most salient developments to facilitate the reemergence of comprehensive rationality. The potential for using quantitative methods for multiple-criteria decision analysis and other tools of operations research in ecological systems is immense. Accessibility to data and learning through electronic means such as the Internet and computerized databases has also transformed the information processing abilities of society, thereby facilitating the development of holistic approaches to industrial management.

Braden Allenby and Thomas Graedel were among the pioneers who first enunciated the principles of industrial ecology in the American corporate realm to harmonize various strands of research into practice. While employed at American Telephone and Telegraph (AT&T) in two different divisions, Allenby and Graedel brought environmental reform to the forefront in the corporation. The difference in their academic backgrounds is an important illustration of the breadth industrial ecology aims to cover. Allenby is a lawyer by training and also has a doctorate in engineering. Graedel is an atmospheric chemist who had spent most of his career at Bell Laboratories. Their work has greatly helped in bringing industrial ecology in the limelight of government policy and educational discourse. Allenby and Graedel are authors of a comprehensive textbook on industrial ecology sponsored by AT&T.[18]

For some purists the term "industrial ecology" may seem to be an oxymoron. How could industrial processes, which had distanced humans from their natural environment, be married to the very natural science of ecology? Part of the answer lies in the realization that, whether we liked it or not, industrial processes were inevitably impacting the natural system and vice versa. Humankind had removed itself from managing these latent interactions, because it was too busy managing the production and consumption of goods and services. Second, the permanence of industrial processes began to dawn on scientists and engineers alike. There was no plausible turning back from the lifestyles of comfort and convenience we had come to accept—and thus natural science would inexorably need to accommodate the industrial system.

Robert Ayres, an American physicist based at a business school in France, developed the concept of industrial metabolism, which aims to describe industries as mega-organisms, consuming low-entropy resources to provide for their own sustenance while allowing for growth and reproduction. Ayres argues that the analogy between biological organisms and industrial activities is compelling since "both are examples of self-organizing, dissipate systems in a stable state, far from thermodynamic equilibrium."[19] Although such interdisciplinary engineers as Ayres were considering industrial metabolism, input-output analysts revolutionized how we consider materials flow in our economy and helped to bridge engineering and economics.[20]

Combining concerns about energy usage and entropy leads to another useful concept, called exergy, which is being explored increasingly as a metric of understanding the irreversibility of certain kinds of resource extractions. Originally developed by the great physicist J. Willard Gibbs in 1873, the concept has been refined as physicists, ecologists, and some economists have joined forces to understand resource depletion. The term itself was first used by Slovenian engineer Zoran Rant in 1956. The exergy of a material can be defined by its ability to do useful work in achieving thermal equilibrium with its environment. Although energy can be neither created nor destroyed under normal circumstances, a material's exergy can be destroyed based on an increase in entropy through mixing and dispersal. Both energy and exergy have the same measurement units (joules), but for understanding material usage and sustainability, exergy is a more useful concept to be further developed.[21]

If the industrial sector is to be considered a pseudo-organism, we must also relate that entity to the broader environment in which it subsists: hence the idea of industrial ecosystems. The interrelationships between the environment and industry extend to biotic as well as societal factors. It is no wonder, therefore, that economics, ethics, and anthropology are often invoked in discourse pertaining to industrial ecology. However, the most consequential disciplines in applying these principles are either business administration or public administration.

Blissful Bureaucracy

The Danes take justifiable pride in their social democracy as a model of good government, especially in the environmental arena. The growing genre of "Eurotopia" literature in the United States usually profiles the fantastic bike paths of Copenhagen or the energy-efficient building design of their infrastructure. However, one of the most remarkable success stories of this small Scandinavian country comes from a confluence of waste exchanges between businesses rather than any particular government program. Although regulations played a part in galvanizing the businesses to action, the industrial symbiosis that emerged in the town of Kalundborg (population twenty thousand) is largely a product of a well-configured treasure impulse for industrial wastes. Though the system of industrial cooperation evolved out of necessity over a period of twenty-five years and without a preplanned organization, it has become almost synonymous with the idea of industrial ecology.

A coal power plant, an oil refinery, an industrial enzyme manufacturer, a plasterboard factory, and the city's district heating department are the main components of the industrial park. Heat exchange is central to the efficiency of the system as predicted by the laws of thermodynamics. The power station's waste heat goes to support fish farms, greenhouses, and district heating. However, there are also other material exchanges that are quite innovative. An oil refinery sends its sulfurous byproducts to a sulfuric acid factory, and butane gas, which would otherwise be flared, to a plasterboard factory. Steam from the power plant goes back to the refinery and also to a pharmaceutical company, which in turn sends its organic wastes to a fertilizer manufacturer. The power plant also produces fly-ash waste, which is supplied as gypsum to the plasterboard factory.

Inspired by Kalundborg, other governments are also beginning to consider how such efforts can be institutionalized through eco-industrial parks, which deliberately collocate industries that can engage in such waste exchanges. The British government has taken a lead in also establishing a National Industrial Symbiosis Programme to facilitate such interactions between businesses.[22] The law and the political system need to operate in the same cyclical way in which the environment and its components work. An excellent example of this approach is the Swedish government's decision to inaugurate a Department of Industry and Ecocycles, which brings together technical expertise from various disciplines and thus exemplifies an innovative policy implementation of industrial ecology.[23]

Since industrial ecology has emerged as a synthesis of several different disciplinary efforts, the task of implementing its principles must take place at all conceivable echelons of society. First, it is evident that we need to somehow integrate regulatory systems, so that they do not look at each medium of pollution separately, but try to understand how various input and output channels are interconnected. This does not necessarily mean that we integrate all statutes, but rather that the licensing procedure for pollution control should be integrated, as is the case in many European countries. This saves tremendous transaction costs on separate permitting procedures and also encourages companies to think about synergies in pollution prevention initiatives, such as redirecting the wastes from one process to another usage. However, to further the aims of industrial ecology, interindustry cooperation needs to be facilitated through corporate initiatives as well, such as Responsible Care, created by the American Chemical Manufacturers' Association, and the Global Environmental Management Initiative (by the International Chamber of Commerce), as well as government involvement to ensure that such efforts have meaningful results and are not an attempt at "greenwash." One does, however, need to be careful in knowing the limits of this approach, because a very large scale interdependence of industries can make all vulnerable to a slight malfunction of their symbiotic components. Therefore, we must have appropriate contingency alternatives for "industrially symbiotic" systems that should be part of the engineering designs for such efforts.

A lot of the win-win options of being green have already been taken. Certain choices will have to be made that may be costly in the short run, but the long-term benefits of those decisions (such as using naturally derived pesticides, which are initially more expensive) will be significant enough to warrant investment. A good example of this long-term approach to competitive advantage is the Japanese automobile industry, which anticipated fuel shortages and designed smaller, more efficient cars in the 1960s. The Japanese gained a strong advantage over their American competitors, which did not take into account such contingencies. Appreciating the importance of material cycles will allow corporations and individuals alike to plan for such long-term considerations by providing a means for looking at the interconnections within systems from energy supply to waste removal. It is important to note that merely reducing materials throughput does not necessarily render the system ecologically sustainable. Processes that transform high-entropy wastes into even higher entropy feedstocks can, in fact, inhibit the industrial ecology goal of achieving closed materials cycles. Critics of the approach contend that an effort to holistically consider all the environmental implications of a product or process "would generate a barrage of formidable tradeoffs between uncertain and widely disparate outcomes that must ultimately be reduced to social value judgments."[24] Yet advancements in computational means of information management and prioritization regimes that can also involve communities effectively as part of the planning process can assuage such concerns.[25]

Nobel laureate Robert Solow noted the important attribute of non-renewable natural resources as key indicators of how markets intersect with planning mechanisms. In his essay "The Economics of Resources or the Resources of Economics," Solow presciently stated: "The underlying logic of indicative planning is that some comparison and coordination of the main participants in the market, including the government, could eliminate major errors and resolve much uncertainty. In the case of exhaustible resources, it could have the additional purpose of generating consistent expectations about the distant future."[26]

Industries need to plan such ventures to ensure that they find the separation and reprocessing technologies to refine the wastes so that high-entropy wastes can be converted to low-entropy feedstock. This process would of course require energy usage, and a detailed assessment of costs and benefits would need to be carried out to assess the environmental impact of this extra energy usage.[27]

What matters most is a realization that industry does indeed have effects on ecosystems and shares many similarities of process with them. Hence anthropogenic ecosystems, such as industries, may indeed gain from a more harmonious interaction with nature. To reaffirm the essential linkages between industry and nature, it has even been suggested that we use the term "earth systems ecology" instead of "industrial ecology." Whatever the choice of words, it is important to keep in mind the concepts we need to implement a systems approach to environmental management.

The American statistician W. Edwards Deming studied industrial development in the United States during the 1940s and was keenly aware of the challenges of developing a complex product with thousands of components. Finding connections between various parts of the production process was central to Deming's approach in improving the effectiveness of large industrial systems. He applied some of the integrative lessons from his research to help Japan build its automotive and electronic sector in the aftermath of World War II. Deming proposed several revolutionary concepts to managing industries that are now collectively enshrined in the field of total quality management (TQM). For his work Deming was celebrated in both his homeland and in Japan, which awarded him an Order of the Sacred Treasure in 1960.[28] Total quality management entails looking at all stakeholders involved in the process of production and consumption of a product and working toward continuous improvement. Expanding on this concept in the 1990s, environmental managers suggested adding an "E" to the acronym and expanding its scope to include all environmental flows in the production process. TQEM lays a lot of emphasis on the measurement of performance, continued change, and innovation. Decision making should be data driven and there should be an emphasis on continuous improvement. Design should be geared toward quality and must anticipate problems as opposed to reacting to mistakes. From an environmental standpoint, this may be achieved through management changes, technological improvements, and the establishment of self-correction mechanisms. There is a need for corporations to institute this management mechanism and for government to encourage its establishment, through institutional cooperation. TQEM programs require far greater communication between various departments of a corporation so that environmental concerns can be tackled collectively. For example, the manufacturing and sourcing department needs to coordinate modular design for easy recyclability with the research and development department of a company. Most large American corporations have developed TQEM programs, and there is now even a quarterly journal by the same title, as well as a *Journal of Industrial Ecology*.

Another important manifestation of the principles of industrial ecology is life-cycle analysis (LCA). The concept emerged in product engineering on account of the realization that a product may affect the environment in adverse ways, both before and after it is consumed. The Society for Environmental Toxicology defines LCA as "a process used to evaluate the environmental burdens associated with a product, process or activity across its entire life."[29] It underscores the importance of manufacturing processes and disposal processes that were previously sidelined by the most palpable aspects of consumption. Detailed LCAs are often expensive to conduct, but with computer modeling and increased skills, they may become much more efficient. LCA is a means of environmental accounting that helps to prevent post-facto remediation efforts and offers a means of achieving long-term efficiency. There is now an evolving international standard for life-cycle analysis under the ISO 14000 series of international standards. This standard was promulgated in 1998 and has gained considerable acceptance in the past decade.

Automobile manufacturing is an area where LCA is particularly informative and useful, since so many materials are used in manufacturing, operating, and repairing cars, and even in associated sectors like insurance and financial services (fig. 16). For each component of the vehicle, we can also develop a separate life cycle in comparative terms and consider product choice accordingly. Lester Lave and colleagues, who pioneered some of the earliest studies of this kind, have used life-cycle analysis to compare plastic fuel tanks versus steel in their ecological and economic costs. Surprisingly, despite the easier recyclability of steel, they found that in overall

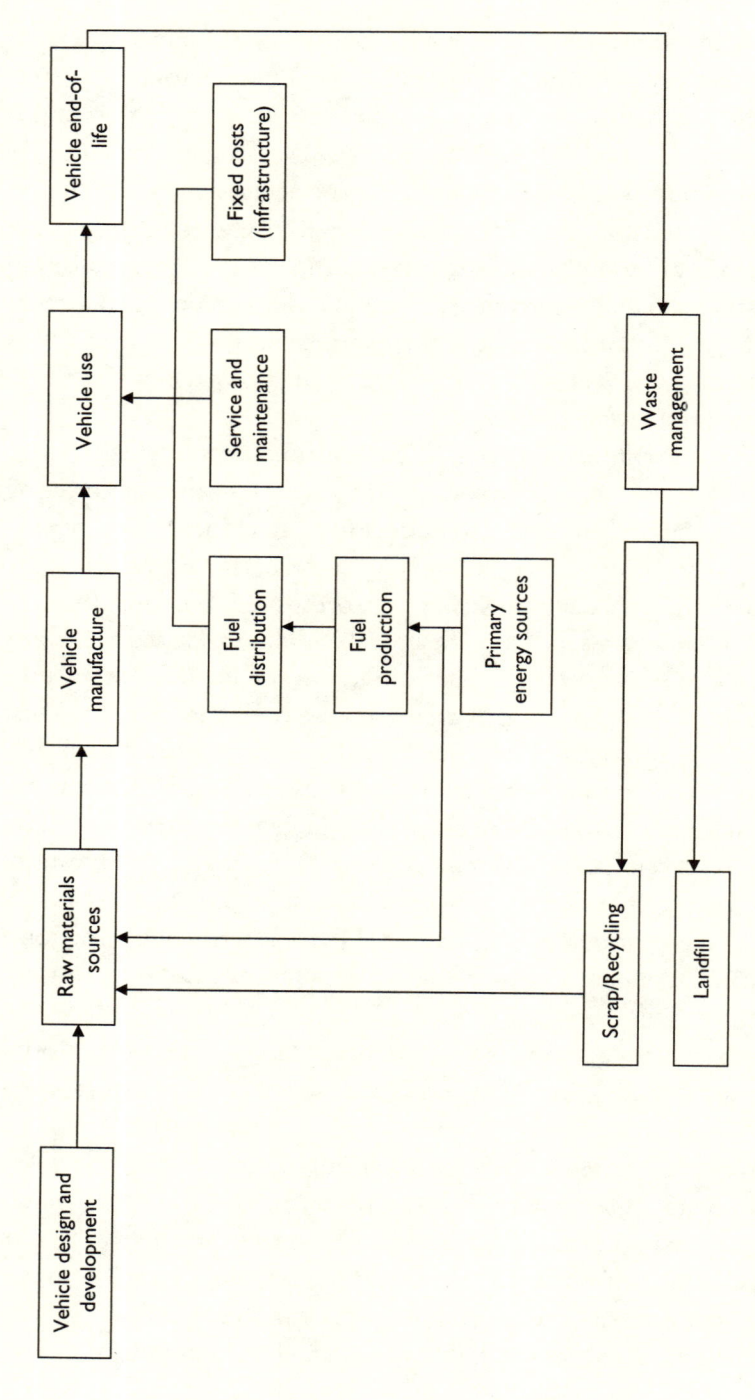

FIGURE 16. Simplified life cycle of an automobile (Heather Maclean and Lester Lave, "Evaluating Automobile Fuel/Propulsion System Technologies," *Progress in Energy and Combustion Science* 29 [2003]: 69)

impact, over the full life cycle of the tanks, plastics were 50 percent more energy efficient and 350 percent less damaging in weighted toxic chemical releases and transfers.[30]

Industrial ecologists believe that a product remains the responsibility of the producer until it is actually absorbed by the system in which it enters. Food products, once they are digested by organisms, no longer remain the responsibility of the producers. However, products containing materials like steel and plastics do indeed remain the responsibility of the producer, since they are not permanently digested or "metabolized" by any entity and become a liability on the system after usage. This is particularly a problem with obsolescent technology items, such as cars, computers, and photocopiers. Increasingly there is a trend to implement organized recycling schemes, where the producer must take back the product after use and is responsible for either refurbishing it or disposing of it in an environmentally appropriate manner. Such schemes have been especially successful with photocopiers and laser printers in the United States.

The Animate Company

Corporations have often been described in organismic terms because of their capacity for impact on the environment and some of their behavioral attributes, such as resource consumption, waste generation, and growth. Corporate unions or mergers are frequently referred to with matrimonial metaphors. Occasionally, corporations may even exhibit a kind of amoebic procreation, as with the "Baby Bells" that emerged from the parent company AT&T in 1984. In their landmark book *Organizational Ecology*, Michael Hannan and John Freeman first expounded on this framing of corporate behavior. There was a Darwinian tone in their analysis that harkens back to competitive markets being akin to natural selection processes. While many corporations may fit some aspects of this model, the growth of the mega-consumer-store, as exemplified by Wal-Mart, challenges this model. When globalization and economies of scale allow for consolidation of all retailing under one roof, the prospect for constructive competition can arguably be stifled. However, the concentration of consumer merchandise by a dominant retailer opens some remarkable opportunities for efficient environmental action.

Policy analysts have discovered that normative issues with future benefits, such as environmental initiatives, are usually most easily implemented through focusing on producers and service providers. Consumers are too capricious and often not adequately informed to even respond to basic market incentives in this regard. Furthermore, the speed of consumer response to technical information or prices might not be able to keep up with ecological stresses. High-income consumers who usually take on a very large share of product consumption might have highly inelastic demand patterns that would not respond to price fluctuations. How, then, might we spur positive change? Where in the food chain of material consumption might we get the biggest bang for the buck?

A few attempts have been made in the management literature to examine various ways of implementing the principles of industrial ecology in the corporate world, most notably in the writings of Paul Shrivastava, who believes that industrial societies have traditionally led managers to focus their efforts on the creation of wealth through technological expansion, whereas managers in postindustrial societies must shift their efforts toward managing risks that accrue from the creation and distribution of wealth.[31] His analysis emanates from a study of major industrial accidents, such as the methyl isocyanate leakage from a Union Carbide plant in Bhopal, India. Risk, he proposes, is the primary motivating force behind the adoption of industrial ecological principles by managers in the modern corporation.

While Shrivastava's model of "ecocentric management" is useful from a macroscopic perspective, he does not offer tangible suggestions for how the business administration of corporations should change. What follows is a brief set of recommendations in nine significant sectors of a modern corporation that would follow the paradigm of industrial ecology. It is important to appreciate that 70 percent of our economy is now in the service sector, and hence many of the changes recommended for the manufacturing sector will inevitably need to be reinforced in the service sector.

1. Strategic Planning Division
    - Plan to integrate vertically rather than horizontally in order to minimize transactional impact on natural resources that are to be used in the production process.

- Avoid large transportation costs and resulting pollution by locating facilities in closer geographic proximity.
- Look for synergies in energy and waste utilization with nearby industries through the establishment of eco-industrial parks. Share best practices with industries to achieve better cooperation.

2. Government and Community Relations Department
   - Lobbying efforts of the industry should focus on long-term voluntary compliance initiatives that could reduce the infrastructure required for compliance-centered government agencies. However, the voluntary compliance must be effectively enforced internally.
   - Favor integrated environmental regulations rather than the conventional air, water, and waste mode of incremental regulations.
   - Dispute resolution and negotiation strategies should be favored over legal action to reduce transaction costs of litigation, unless it is important to set an institutional precedent with a case.

3. Research and Development Division
   - Utilize industrial ecology concepts of Design for Environment and Dematerialization to develop more eco-friendly products.

4. Manufacturing/Sourcing Division
   - Coordinate activities with R&D sector to ensure that manufacturing processes optimize energy usage in making products.
   - Choose suppliers that are in close proximity to the manufacturing location to reduce transportation costs and risks of environmental accidents.

5. Environmental Health and Safety (EHS) Department
   - Move from compliance-oriented EHS management to proactive pollution prevention.
   - Work with R&D department to see if emissions can somehow be reused in the manufacturing process in your company or in other nearby companies.

6. Financial Management and Accounting Departments
   - Use a low discount rate for evaluating the future benefits of environmental projects in order to ensure that the long-term benefits are accounted for. Consider reporting performance over longer time horizons as well.
   - Include the cost of resource depletion or resource amelioration when calculating a company's profitability.

7. Human Resources Department
   - Provide training for all employees on environmental issues so that company initiatives are appreciated and taken in context (an ecoliteracy requirement across the company).
   - Encourage employees to live near the company's location and provide incentives for use of collective or public transportation.
8. Marketing Department
   - Reduce advertising through paper-based or other disposable media.
   - Persuade industry competitors to produce and market items with ecological impact data and benchmarks on product labels to encourage positive competition on environmental performance.
9. Customer Service Department
   - Encourage customers to participate in product-takeback by offering incentives for recycling and return of products.
   - Provide services for product repair or telephone guidance for home repairs, with modest fees as a revenue stream rather than encouraging obsolescence.

The measures described here are by no means exhaustive, but rather provide a skeletal set of points to consider in the context of reconfiguring a corporation to account for material cycles. These measures also attempt to keep in view the fact that manufacturing does indeed provide lots of jobs but that services associated with product repair and recycling can provide comparable impacts, which can be evaluated through life-cycle analysis. Efficiency of resource use and dematerialization are attractive propositions that appear to be win-win outcomes. However, there has been some concern about their application in the larger context of environmental policy prescriptions on natural resource usage. The empirical evidence on whether efficiency can actually moderate consumption has been widely debated since the advent of the industrial revolution. British economist William Stanley Jevons published a book titled *The Coal Question* in 1865 in which he presented data that suggested that increased efficiency through technological advancement was actually leading to greater consumption. The explanation he proposed was that, since efficiency is likely to reduce cost, the result will be greater demand for the same good, and a "rebound" in consumption will follow. This observation is known as the

Jevons paradox, and is often presented by environmentalists as an antidote to technological optimism with regard to energy and material efficiency. The same argument can also be made for dematerialization—as less material is used for manufacturing a product, more of the raw material will be available for other goods, spurring latent demand in other products that could use that material.

While the paradox may well hold in the case of materials and energy usage areas with growing demand, it is less likely to be valid in mature markets. For example, in a developing economy energy efficiency and reduced cost may lead people to buy more products to use more energy. However, in a market with a fairly saturated consumption profile (not many more things to add to material and energy demand), efficiency and dematerialization are very valuable. We can thus be cautiously optimistic about the prospect of channelizing our treasure impulse toward seeking efficient material usage.[32]

Critics of industrial ecology summarily argue that structured management leads to centralized planning, which has failed throughout history, starting with the Greeks and Romans, and more recently with the Soviet Union. It is important not to confuse thought and process at this juncture. The vision of industrial ecology, though stressing the intrinsic interrelationships between various biotic and abiotic processes, does not necessarily imply that we centralize efforts to deal with all material problems. We need to be far more discerning than to simply propose a single set of solutions. Where do we draw the line between integrating holistic ideals (that are implicit in a systems approach) on the one side and practical implementation in segmented form on the other? The best way to approach this question is to first try the integrative or holistic approach at the communication and networking level, without instituting widespread organizational change. Once successful interdepartmental and interdisciplinary programs have been implemented, then organizational change to integrate synergistic systems may be implemented more effectively as well.

Tying together material cycles across spatial dimensions of resource availability and waste management schemes can be achieved through the paradigm of industrial ecology. However, we are still left to consider the temporal dimension of resource extraction. Planned conservation and efficient cycling of wastes may extend our time horizon for depletion and perhaps give technology a greater opportunity to find alternatives. However, we must still contend with the ecological impacts at each phase of extraction, particularly at the primary level when ecosystems are most acutely disturbed. Each extraction location will eventually need to be restored to some level of functionality if we are to consider any potential for sustainability at the planetary scale.

8

THE RESTORATION ETHIC: HURTING AND HEALING ECOSYSTEMS

> The nation behaves well if it treats the natural resources as assets which it must turn over to the next generation, increased and not impaired in value.
> —Theodore Roosevelt, *The New Nationalism,* 1910

The Pecos River might not be familiar to most Americans as a notable waterway, but for the residents of New Mexico and western Texas it has been a perennial presence for generations. Trout from the river have provided an abundant source of protein for Native Americans and settlers alike. Water from the Pecos was the lifeblood of the Carlsbad Reclamation Project, one of the earliest irrigation ventures of its kind, initiated by the federal government in 1902.[1] This sparsely populated and scenic region of the American West was home to none other than Sheriff Pat Garrett, known to Wild West enthusiasts as the man who shot dead the infamous outlaw Billy the Kid in 1881. Soon thereafter, Garrett established an eighteen-hundred-acre farm near the town of Roswell, long before it became a mecca for alien hunters and fans of *The X-Files*. The aim was to develop the ranch for agricultural purposes by "reclaiming" the water of the Pecos River.

Those were the days when the connotation of reclamation was to harness as much of a resource as possible lest it be "wasted" through natural processes. In the case of water, allowing it to flow into the ocean without being used productively was considered wastage, whereas utilizing it for irrigation was considered "conservation." A century later, as humanity attempts to grapple with the challenges of ecological decline, the meanings of both conservation and reclamation have changed dramatically.

When Janice Varela, a resident of the village of Pecos, New Mexico, first contacted the fisheries management group Trout Unlimited, in the mid-1990s, she was interested in reclaiming the Pecos River for quite a different purpose. Fish in her beloved river were perishing mysteriously, and she was determined to reclaim the quality of the water from whatever was afflicting it. Janice's fact-finding mission led her to various government agencies and environmental labs and finally to the Cyprus Amax Corporation, which had operated a zinc-mining operation upstream several decades earlier. Finding the company's Terrero Mine as the source of heavy metal pollution was a shock for Janice, as she was not there to go after deep pockets for damages at all. Much of her community remembered having decent relations with the company during the mine's operative years, from 1930 to 1965, and there was a "reluctance to admit risk," even though the site had been declared a Superfund cleanup site by the U.S. Environmental Protection Agency.[2]

Yet people in the community knew that no matter how much they had valued the jobs of the company historically, their ecosystem needed to be given priority, so they set out to find a way to restore the riparian environment.[3] Fortunately, the company did not hide behind lawyers or technicalities about proving impact, and set out to find a solution to the problem. Waste rock that had found its way into the river and was slowly leaching metals had to be physically removed. A range of analytical techniques, such as x-ray fluorescence, was used for soil and tailings characterization for heavy metals, particularly lead. Controlling the off-site movement of tailings was a priority of the project, and this was accomplished by repairing an embankment and constructing a diversion channel to carry surface water above the tailings. Concrete cutoff trenches, dug into excavated bedrock, further strengthened the system along with a liner that provided a barrier to subsurface flow. Not all efforts at fixing the problem achieved immediate success. A liner intended to prevent contaminated water from seeping into the groundwater started to leak within months of installation. However, the team remained on the job for the next several years to ensure that all such issues were addressed and the river was indeed "restored" to a functional level that would support a healthy ecosystem.[4]

The restoration process at the Terrero Mine may be considered a relative success story of how a degraded landscape was recovered through engineering solutions and community commitment to its rehabilitation (fig. 17). Not all mining areas are able to claim the same level of success, however, and the scale of past extractive enterprises across the planet is staggering. Data on abandoned mines and drilling projects around the world are hard to come by, but in the United States there are around 557,650 abandoned mines scattered over thirty-two states, together accounting for 50 billion tons of untreated waste rocks. Mine wastes have polluted 180,000 acres of lakes and reservoirs and 12,000 miles of streams.[5] Cleanup of the sites listed under the Superfund law by the federal government alone is likely to cost around $24 billion. Many of these sites are so-called megasites where mining has occurred over decades, and the effects are often not visible to visitors but are quite apparent to residents and wildlife. Idaho's Coeur d'Alene Basin, which is usually associated with postcard views of mountain valleys and trout-filled streams, is an example of such a megasite, which constitutes a 150-mile corridor that has absorbed more than 118 million tons of persistent metal mining wastes in its history.[6] A separate study of currently operating metallic mines in the western United States in 2001 revealed that, because mining companies may have inadequate reclamation bonds, cleanup costs of around $12 billion could end up being charged to taxpayers.[7]

There are many variables that need to be considered in such ventures that go beyond the visible footprint of a mine. Consider the predicament of the Canadian government in dealing with the aptly named Giant Mine reclamation effort in the Arctic reaches of the Northwest Territories, near the city of Yellowknife. Contaminated wastes containing arsenic trioxide from the mine had seeped into the groundwater system, but were immobilized through natural processes of permafrost. However, the melting of permafrost layers in this region in the past decade or so as a result of climate change has resulted in greater chances that the toxic effluent will be mobilized again through the groundwater system. To reclaim the land and prevent the fouling of the ecology in this region, the only practical solution the Canadian government could find was to emulate natural processes that had thus far kept the toxins on a leash: it is attempting to "generate" permafrost by injecting cooling rods underground to refrigerate the toxins and keep

FIGURE 17. Restored Terrero Mine, New Mexico, showing how the old mill buildings have been capped and adjoining areas revegetated, 2004 (New Mexico Energy, Minerals, and Natural Resources Department)

them from moving into water supplies. The reclamation effort is an ongoing energy drain that could remain in perpetuity, or at least until a more cost-effective method can be found for immobilizing or extracting the pollutants for productive use.[8]

The task of dealing with past ecological stresses from the extractive sector is clearly a herculean one not only for North America but across the world. Yet here, too, the treasure impulse is finding ways to reclaim fortune from what might otherwise have been a huge hole in the ground.

Caledonian Quandary

The Romans came to the British Isles with visions of conquest but were eventually repelled by the northern tribes, particularly a Pictish group known as the Caledonii. Roman narratives therefore refer to present-day Scotland as Caledonia. Much of the landscape of this region, with mountains and deep ravines, favored the development of tribal cultures that fiercely defended their territory. The northern region of Scotland, known as the Highlands, eventually became drawn into the social, political, and economic influence of the rest of Great Britain. Many of the former Highland chiefs, who often retained title to large tracts of land, took on new roles as landlords, or lairds, and their former clansmen as tenant farmers. The success of this transition was helped by the importation of sheep to the region, as the rigors of the climate and geography would allow for little else in the way of large-scale agricultural development. With the price of wool soaring at the onset of the nineteenth century, sheep were perceived as more valuable than people, which led to the wholesale removal of tenants from the land in an era known as the Highland Clearances. Many of these former tenants relocated along the coastal areas of the Highlands and offshore islands on less productive lands, where they settled on small tenanted agricultural holdings referred to as crofts. Crofters remained indebted to the landowners and were often strapped financially by increasing rents. This necessitated supplementing their meager agricultural wages by fishing in coastal waters and harvesting kelp, which was burned and converted to alkali for the expanding industries of the south.

As a consequence of this land-use history, Scotland is renowned for having an extremely skewed land ownership pattern. In the latter part of the nineteenth century, a mere 118 people owned half of Scotland, with much of the land accumulated into large sporting estates. Today, this figure has increased to over 600 landowners, but Scotland still has the most concentrated pattern of private land ownership of any region in the world. The situation is even more extreme in the northern mountain and western island regions, collectively referred to as the Highlands and Islands, where 340 sporting estates account for over half of the privately owned land.[9]

As large hunting estates that might be managed as "conservation properties" become more expensive to sustain, they are likely to be sold for development. In remote parts of Scotland, the market for such sales would be highly limited because the properties are difficult to reach and there is little economic development nearby. The mining and quarrying industry is one sector that might consider remoteness an attractive feature for development, because opposition would be less likely in such areas and restoration might be more efficiently carried out.

The Morvern peninsula, spanning some 94,000 acres on the remote west coast of Scotland, in what used to be Argyll and is now part of the Highlands jurisdiction, is one such region that has limited road access and is one of the least populated parts of the mainland. The entire region has a population of only around eight hundred people. Easy access to a deep waterway also makes this an ideal place for a quarry that would export heavy rock material. Coastal quarries were first publicly considered in Great Britain in the Verney Report, published in 1976, which identified such large remote operations as a viable supply option.[10] The British government had also been trying to move from land-won sources of aggregate to marine sources and coastal super-quarries in remote locations that provide an economically more attractive compromise.[11]

Aggregate tycoon John Yeoman, the son of Foster Yeoman who founded one of Britain's largest aggregate mining companies, had been a strong proponent of a super-quarry to be situated in a remote location from which stone could be exported by sea. Yeoman acquired property on the Morvern peninsula in 1982 without much dissent at the time. The land had previously comprised two highland hunting estates, Glensanda (5,900 acres) and Kingairloch (14,000 acres). The negotiations to develop the quarry also led to the passage of the Glensanda Harbour Act, which provides for harbor operations in a leased area of about seven square miles in Loch Linnhe.[12] Public access to the quarry is possible only by a half-hour boat ride across the loch. The environmental and social review for the Glensanda quarry took about two years, and permission for development was granted in 1986. Under Scottish law, mineral rights belong to the landowner except for gold, petroleum products, and coal — so the mineral rights to the aggregate automatically transferred to the buyer.

In some ways this was an ideal situation for a land transfer from an estate to a private operator, since there were few competing interests and environmental groups had not been galvanized to action in this location. The quarry itself is not visible from eye level on the other side of Loch Linnhe, which may have also accounted for the lack of resistance. Permission to quarry was granted for a period of twenty-five years, and Glensanda is now one of the largest quarry operations in the world. It has in excess of 150 years' worth of reserves of granite, or more than 900 million tons. The operation currently produces 6 million tons per year, and is designed to be able to produce up to 15 million. The quarry operates 24 hours a day, 365 days per year.

The owners of Glensanda have agreed to multiple uses on the property, including deer hunting, and part of the Kingairloch estate continues to offer recreational accommodations. In addition, the quarry is attempting to create an artificial reef through the placement of waste rock in Loch Linnhe, because tourists who visit for underwater diving are an important economic resource in the nearby village of Lochaline. In contrast to the campaign against a similar quarry proposed on the Isle of Harris, in which fishermen joined with environmentalists, the Glensanda case did not lead to opposing alliances. One of the most strident opponents of the Isle of Harris quarry, Alastair McIntosh, acknowledges that if there is a need for a super-quarry (which environmentalists argue is questionable due to stagnant demand), Glensanda is the appropriate site for it, primarily because of its potential for ecological restoration.[13]

There are even plans to restore some of the physical infrastructure for other uses once the mining stops. The stone is initially removed through conventional large-scale quarrying methods, including drilling and blasting, loading and haulage, and primary crushing operations, all carried out about 1,500 feet above sea level. The primary crusher feeds a vertical shaft known as the "Glory Hole," which transports the crushed rock down to sea level without conveyors, and hence minimal visual impact. This technique has so far avoided any large-scale dust accumulation on the side of the property that can be seen from Lismore. Although parts of the larger pit are visible from Lochaline, and are also a source of light and noise pollution, the opposition to the quarry has been minimal.

Further expansion of the quarry is being proposed, and the environmentalist challenge is being raised in simple terms: preservation of a relatively wild landscape over industrial development. The granite being mined is a nonrenewable resource, so the key issue is that, even though the life of the quarry could possibly be extended for as much as 150 years, there will ultimately be a further transition when the mining eventually stops. Because previous use of the area was relatively limited to wildlife and recreation, the restoration of the landscape is an essential element of the development. The restoration plan for the site aims to "integrate the production benches with the surrounding hillside." The company is sequentially restoring terraces, starting with the more visible upper benches. Rock is blasted down to form screes, which are left at the natural angle of repose. Peat is stripped and used directly in restoration by pressing it into the rock face or scree to help it adhere and prevent it from being washed off. Eventually, the site will simulate the surrounding landscape and be returned to moorland. According to a Scottish government planning document, eventually the "Glory Hole" shaft may be developed into a pump-storage hydroelectric scheme. The company has also tried to support local aquaculture as an alternative livelihood and is planning to build wind turbines on the property.[14]

The Glensanda operation has thus far been able to concurrently operate with alternative livelihoods of tourism, diving, and some limited fishing in the area. Opponents of the quarry can only point toward preservation of the landscape as the alternative land use. However, as depopulation continues to be a concern among rural Highland communities, the transition to environmentally moderated industrial uses such as Glensanda appears to be a viable option.

More than twelve thousand miles from Glensanda, in the Coral Sea of the South Pacific, is a small island that reminded Captain James Cook of Scotland when he first sailed around it in 1774. Conifer trees reminiscent of Scotch pines were perhaps the feature that caught his attention the most, and he named the island New Caledonia. The island boasts one of the highest rates of endemism in the world: an astounding 90 percent of the species that exist there are not found naturally anywhere else. In *The Diversity of Life*, the naturalist E. O. Wilson called New Caledonia his "favorite island."[15] The fortune of this tropical land would, however, be linked not to some exotic appeal of its flora or fauna but rather to its subterranean wealth in the metal nickel. The island is home to one of only three major nickel deposits in the world (the other two being in Sudbury, Canada, and Norilsk, Russia). New Caledonia became a colony of France in 1853, and nickel mining has been the mainstay of its economy ever since.

Around 45 percent of the island's population are indigenous Melanesians, sometimes known as Kanaks; 35 percent are of European descent, 10 percent are of Polynesian origin, and the remainder are a mixture of Vietnamese, Chinese, and other Asian ethnic groups. Responsibility for environmental management falls almost exclusively to the local provincial governments. As an overseas territory of France, New Caledonia is not subject to French laws and regulations unless these are adopted by the local territorial government. The provinces are free to adopt the whole or any part of French environmental legislation, but so far they have not opted to do so. No restrictions of any kind existed until about fifteen years ago, which allowed nickel to be recklessly extracted for decades. Much of the island's landscape has been scarred by mining or logging, leading to widespread erosion and excessive siltation in rivers, which is also beginning to affect the coral reef surrounding the island—the second largest in the world.[16]

There is no national restoration plan to repair the damage from mining. Nongovernmental organizations have made some efforts through corporate partnerships that have received mixed receptions from the local population. In 2003, the World Wide Fund for Nature (also known as the World Wildlife Fund, or WWF) entered into a collaboration with the French industrial materials company LaFarge to develop a dryland forest restoration program in New Caledonia. This is part of a larger partnership between the two organizations (one of many such arrangements the WWF has initiated with international corporations), in which LaFarge has committed $1.1 million per year over five years to WWF programs. This collaboration has been largely well received in the areas where it is operating. However, when WWF-Canada tried to engage in a similar partnership with the Inco nickel-mining company of Canada, including a large financial contribution for long-term landscape restoration, the public reacted with outrage over the perceived corporate cooptation of environmental interests.[17]

Reinventing the Landscape

Adaptability has been a key feature of human societies, allowing inhabitation of some of the warmest and coldest places on the planet with a good measure of tenacity. Complaining or celebrating, humans adjust to their surroundings and build societies accordingly. Nevertheless, once they have been there long enough, people also form attachments to places—every contour of the landscape can have subjective significance to each of us. Altering the landscape is acceptable at times but needs to be negotiated with great care so that those who will long for the past may feel some sense of redemption.

It was in this spirit, with mildly theological undertones, that an old clay mine in Cornwall, England, was "regenerated" into what is now called the Eden Project. Mined for metal and industrial minerals like clay, the Cornish landscape has been altered by humanity for centuries, and not until the turn of the twenty-first century has there been a concerted effort to rehabilitate the damage. Pits of china clay have dominated the mid-Cornwall region, and the wastes generated from extraction have been immense, so much that artificial hills have been created over vast expanses of the region and are called the "Cornish Alps."[18] The goal in the Eden Project was not only to build a museum in tribute to the mining heritage of the area but also to celebrate the prospects for a sustainable future. Tim Smit, the founder of the project, wanted it to be a productive economic enterprise, coinciding as it did with the Millennium Commission's development efforts in Britain in 2000. Unlike the ill-fated Millennium Dome, which became insolvent in 2003, the Eden Project flourishes today despite occasional rumblings of discontent about excessive traffic and congestion from the influx of tourists to this remote rural area.

A twelve-acre gray pit that was a blight to any eye was the property with which the designers had to work. Clay pits do not usually have to contend with heavy-metal pollution concerns, nor do they have the impending threat of acidic drainage, which is a formidable challenge to remediation efforts for metal and coal mines. The easiest reclamation idea would have been to vegetate the area and create a sort of sunken garden with perhaps a lake in the deepest portions of the cavity. One of the world's finest gardens, the Butchart in Victoria, Canada, had the pedigree of a former quarry as well. An earlier mine reclamation in Cornwall had followed a similar path by creating the Lost Garden of Heligan a few years earlier. The Bodiva clay pit at the Eden Project would have made for an interesting garden excursion, but generating large-scale tourist revenues from such a reclamation needed more than just a garden. However, the theme of renewability was vitally important to the designers of the project, and plants were central to any process of terrestrial biotic renewal.

Botany and engineering have a natural convergence in greenhouses, so the Eden Project designers set out to construct the world's largest greenhouse in the Bodiva pit. The result is a series of huge domes constructed of

FIGURE 18. The Eden Project, Cornwall, England (Photo by Kristian Maynard, Bristol, England, used with permission)

hexagonal and pentagonal plastic panels, which together appear as symbolic eggs being hatched from the mine (fig. 18). The greenhouses are reminiscent of the geodesic domes designed by Buckminster Fuller, much of whose erratic but productive life was devoted to understanding the chances for human survival with constrained resources and how to engineer a way out of intractable environmental challenges. He is known for coining the term "Spaceship Earth," which became an evocative metaphor for the environmental movement.[19] Fuller also designed the United States Pavilion at the Montreal Expo of 1967, which now stands as a kind of greenhouse museum operated by Environment Canada.

Using greenhouses to cultivate exotic plants can have an important role in restoration research itself. Often, selective plants, fungi, and their symbiotic bacteria can play a cleansing role in ecosystems depending on the kinds of pollutants involved. For sites contaminated with metals, some plants, such as those of the genus *Cnidoscolus* from Brazil, are "attracted" to metals and can metabolize elements like nickel.[20] Growing such plants for experiments in restoration efforts at a facility like the Eden Project with a clear mandate for post-mining research is a productive enterprise on its own.

Keeping the pit stable with all this engineering being undertaken required a major investment in slope stabilization on the walls of the cavity, which took two years of planning. The designers also wanted to show how the eighty-five thousand tons of topsoil needed for the greenhouse plants could be locally produced, so they set up a partnership with soil scientists to create a suitable recipe using the waste materials found in the pit. This ecologically responsible approach also saved the builders thousands of dollars in production and transportation costs for topsoil. Architect Nicholas Grimshaw has been quite intent in developing the structures of the project to mimic nature in various ways. The latest addition to the project is a copper-roofed building called the Core, which has been designed geometrically using the botanic principles of phyllotaxis. The biomimicry of the design follows a mathematical structure that most plants use for their growth, easily seen in the opposing spiral structures in sunflower heads, pineapples, and pine cones. The copper used for the building is from traceable mine sources where mining is conducted in accordance with criteria set forth by various environmental and social advisers to the project.

A few hundred miles north of the Eden Project, another quarry was rehabilitated around the same time but with a rather different purpose in mind. The steep chasm of any quarry can be an attractive feature for rock climbers, so the purchasers of the Ratho quarry near Edinburgh decided to build the world's largest indoor rock-climbing center in the pit. Not much of the rock in the quarry itself was of suitable quality for climbing, but the structural features of the pit made it easier to build a series of climbing walls at a cost of around $50 million. The designers, led by architect David Taylor, were clearly interested in blending the purpose of the center with the natural aesthetic of the quarry itself. However, the road to success has been rather rocky for the Ratho Center. Two years after opening, it ran into financial difficulties because of modest demand for its services. Although the quarry had been placed in that spot for geological reasons, the center might have fared better in a more accessible suburb of Edinburgh. However, with adequate marketing and a bailout from the

Edinburgh City Council, the center is back on track to host some events at the Olympics in 2012.[21]

Blending old mines with the existing landscape and livelihood regimen is essential for an effective restoration effort. In the relatively arid Indian state of Rajasthan, restoration of ancient mines is being undertaken with a clear recognition of the existing ecological uses of the landscape through traditional water harvesting systems known as *khadin*. This approach utilizes the topography to create an impoundment in small rocky valleys, with a sluice system for discharging any standing water to crops when needed. Old mine pits can be redesigned to create similar arrangements in proximity to local crop fields, thus providing communities with a service they are already accustomed to using. The nearby Sariska tiger reserve is also being integrated into the remediation effort through a revegetation program that would create habitat for the fauna and reduce their chances of coming into human contact in populated areas.[22]

The sinister connotation of being entombed in a mine rings especially true for the multiple uses of the Simsbury copper mine in East Granby, Connecticut, in the United States. For fifty-four years the mine was used as a prison, and for the earlier years of the project, the prisoners were also used as mine labor. The old copper mine had operated since 1705, but ore was not processed on the site. Two years after commencing operations, the proprietors entered into a contract with a group of clergymen, who agreed to operate a refinery and cast the metal into bars for transportation to market. After deducting one-tenth of the revenues for the town (of which two-thirds was to be allocated for the schoolmaster in Simsbury, and the other third to Yale College), the remainder was to be equally divided between them and the proprietors, or workers of the mine. In 1709, the colonial legislature passed an act vesting the right to control all matters relating to the mine in the major part of the proprietors, according to the interests of each. It was under arrangement with this group that milling operations were carried out, until Connecticut began to use the mine as a prison in 1773.[23]

In areas of large-scale mining over generations, the physical changes in the landscape are inextricably linked to livelihoods. Among all mining sectors, coal tends to have the longest duration for mine activity because of its relative abundance, and communities in coal-mining areas, such as

Appalachia, require the greatest assistance in restoration efforts at both the physical and social levels.

Reclamation of coal mines is regulated under a specific law enacted by the U.S. Congress in 1977 called the Surface Mining Control and Reclamation Act, which created a government agency to regulate this sector. The matter at the time was immensely political, since President Gerald R. Ford had previously vetoed bills to enact this law. Jimmy Carter's campaign in 1976 made a pledge to move forward with the law if he was elected, which he did soon thereafter. The Office of Surface Mining Control and Reclamation has operated since then to regulate the industry, with its most significant challenge being to oversee the reclamation of abandoned coal mines where operations had ceased before the enactment of the law. The legislation also created a separate Abandoned Mine Land fund to pay for the cleanup of such mines, financed by a tax of thirty-five cents per ton for surface-mined coal, fifteen cents per ton for coal mined underground, and ten cents per ton for lignite.

The restored landscape should be suitable for use in alternative economies. Although tourism is often the most common alternative presented, many communities may consider more physical uses of the land to be a better option. Many Appalachian communities are interested in developing the cultivation of ginseng plants after ecological restoration of mining areas. Given the rise in demand for American ginseng (*Panax quinquefolius*) in the herbal market, this could clearly be a valuable commodity for a post-mining economy. Climatically, the region is well-suited for the species, and there could be a harvesting regimen that is monitored for sustainability with state and federal forest authorities. However, recent research on wild ginseng in the Appalachian region has shown that the species has declined in its ecological viability, possibly due to environmental impacts from mining activities.[24] Restoration efforts therefore cannot simply revegetate a mine, but rather have to carefully calibrate the future productive use of the landscape, particularly in watersheds, which are often a litmus test for the overall ecological functionality of the region.[25] Considering all these variables in designing an effective remediation plan for a site needs resources of its own. For an obsolescent sector, funding can be motivated only through regulation, since there is absolutely no market incentive for an extractive operator to clean up on its own.

Resourcing Recovery

Australia's aboriginal peoples recollect their traditional conception of property rights in the context of extractive resources through various oral histories. One account recorded in the Melbourne Museum tells the story of Willamimurring, a hill in New South Wales, which was a source of greenstone ax-heads for the Woi Wurrung people. Until his death in 1846, an elder named Billibellary held rights to the quarry that all of the adjoining tribes respected. A European chronicler of the time, William Barak, recorded in 1882 that "when neighboring tribes wanted stones for tomahawks they usually sent a messenger to Billibellary—three bits of stone were given for a possum rug."[26] Similar accounts can be found throughout the continent of various bartering systems recognizing that extraction of a resource had a cost of recovery that even hunting and gathering societies realized at various levels. At small scales the physical environment could be left to revitalize itself, but as the scale of extraction increased, direct intervention would be needed to recover the land. Finding adequate resources to undertake such recovery, particularly in desolate areas with few external incentives for investment, is daunting.

The lonely stretch of asphalt that extends six hundred miles between the pearling town of Broome and the coastal farming community of Kununarra in Western Australia might well pass for a movie set from the lawless age of mineral exploration. Indeed, *Australia* (2008), starring Nicole Kidman, about nineteenth-century life in the outback, was filmed in this region. Yet despite the desolate expanse of the landscape and the paucity of police cars or other indications of regulatory authority, this region has been contested through courts and regulatory regimes. The road is punctuated with signs for extractive enterprises at various levels of development, along with roadkill of kangaroos in higher density than anywhere else in the country. Dirt roads with well-marked tire treads usually follow such signs, but seldom does one actually see anything noticeable beyond the groves of baobab trees and the profusion of red-earth termite mounds. Aboriginal people of this area call the trees larrkardiy and the termite mounds jilkarr; collectively, they hold spiritual significance for these communities and were used in survival strategies, and hence are protected under the country's Aboriginal Heritage Act of 1972.

This remote part of the continent is known as the Kimberley because it shares a common fortune with the region of South Africa of the same name. A third of the world's diamonds come from the Argyle diamond mine here, where the land is claimed by the Mirriuwung, Gidja, Malgnin, and Woolah aboriginal communities. Each of these groups has associations with the land in terms of their ancestral lineage, traditions, and ecological attachment with particular plants and animals. Initially, only a "good neighbor" agreement was established with the aboriginal communities. However, as the size of the mine grew and concerns were raised at various regulatory levels about the mine, a more detailed agreement that would consider issues of post-mining land rights for the communities began to seem necessary. In 2003, Argyle Diamonds acknowledged that past practices had violated trust with the aboriginal people, made an unconditional apology, and offered financial reparations of $1 million to the aboriginal "traditional owners" for disturbing a significant site known as the Barramundi Gap (fig. 19).[27]

In 2005, the Argyle Participation Agreement was negotiated between the Kimberley Land Council, representing the indigenous communities, and Argyle Diamonds, the owner of the mine. Argyle Diamonds agreed to hold the grazing lease in trust for traditional owners for the life of the mining operation, at the end of which the lease will be transferred to them. This transfer would in essence enable the traditional owners to lodge a claim for full-strength native title to the grazing lease area, which Argyle agreed to support. In the meantime, any piecemeal reclamation of the land to meet environmental regulatory requirements would be undertaken in consultation with the aboriginal communities and use native vegetation. This was a historic agreement in that the company agreed to it voluntarily, but the question of actual recovery cost of the land remains unclear. Argyle Diamonds has committed to provide training in land management and rehabilitation as part of the negotiations in the agreement. The mine site includes major infrastructure, including the airstrip and Argyle village, and traditional owners will have an opportunity to submit a business plan to maintain and utilize the infrastructure after the closure of the mine.[28] Financing the recovery of such projects, even those for relatively responsible mining companies, is still a highly contested area. Trust funds and security bonds for reclamation are usually required at some level in most developed

FIGURE 19. The Argyle diamond mine pit, Western Australia, looking in the direction of the Barramundi Gap

countries, but the calculations for these costs are usually very questionable. In one study conducted by the World Bank to consider mining royalty arrangements and their impact on project cash flows, reclamation expenses were largely excluded in the modeling assumptions. The authors of the model acknowledged that "no allocation has been made in the early years for the escrowing of funds to satisfy this future liability . . . this issue may be economically substantial."[29]

Reclamation security bonds in all major developed mining economies have been woefully inadequate. A prime example is the aforementioned Giant Mine near Yellowknife, Canada: in 1999, the mine's owner, Royal Oak Corporation, declared bankruptcy and the Canadian Department of Indian Affairs and Northern Development was left with a $293 million liability. The reclamation security bond for the mine was only worth a paltry $400,000.[30] Instituting mechanisms for ensuring adequate expenditures for recovery is essential in calculating the cost of a new project and the revenue management regime. Unlike other industries, where multinationals can always try to move offshore to avoid more stringent compliance costs for such measures, location of mines is geologically determined, and it is less likely that such a "race to the bottom" would occur.[31] Nevertheless, international agreements between countries that set forth minimal measures of reclamation security should also be negotiated to ensure environmental justice across mineral economies worldwide. Thus far, most of the efforts at global standards for responsible mining have largely been at a voluntary level or through some institutional oversight when financing from donors such as the World Bank is involved.

Conservation efforts are often a major challenge when communities are at the cusp of transition from one source of livelihood to another. However, such transitions can also be opportunities for innovation that can positively shift the development trajectory of communities toward sustainability. Allocating sufficient resources for the economic and social recovery of a mining community requires the creation of further incentives for investment from a different kind of treasure seeker.

Seven miles north of the U.S.-Mexican border lies the town of Bisbee, Arizona, established in 1880 as a mining community around the Copper Queen Mine. By 1902, the town was the largest settlement in the United States between St. Louis and San Francisco, with a population of around twenty thousand. More than 8 billion pounds of copper, 102 million ounces of silver, and 2.8 million ounces of gold were extracted from the Bisbee mines in the following seventy years. When the mines finally closed in 1974, the community began to consider how to sustain a settlement that had declined in population to only a few hundred individuals. The town's proximity to the border was a mixed blessing. It could potentially attract businesses that have facilities in Mexico and needed a bedroom community on the U.S. side of the border, or it could also be a deterrent to investors, given the encumbrance of border control activity against illegal migrants. The nearest city of some size, Tucson, is over ninety miles away and not a reasonable commuting distance for the town to reinvent itself as simply an extended suburb.

To survive, Bisbee needed a plan for alternative livelihoods. It had to generate a tax base that could provide for services for the community and be self-sufficient. Carefully calibrated government programs to create the right incentives for a new breed of treasure seekers are essential in this context. In the case of Bisbee, such help came in the form of the Arizona Enterprise Zone Program, which was enacted in 1989 with the goal of improving economies in areas of the state with high poverty and unemployment rates. The program provides income-tax credits for nonretail businesses that can create new jobs. Property-tax credits for businesses, including selected retail enterprises, were also added to the program in 1993 along with insurance subsidies for employers that went into effect in 1999. As a result, the once atrophying town is now home to seventeen antique stores, twenty restaurants, three museums, thirteen parks, an eighteen-hole golf course, and numerous ancillary businesses. The population growth rate is a manageable 1.4 percent up to 2010, and unemployment has decreased from more than 10 percent in 1996 to just over 4 percent in 2007. Downtown Bisbee reflects the recovery of the town economically, even though the gaping lavender pit of the Copper Queen Mine is still far from ecologically restored. Tours of the mine are a popular draw as well and certainly augment the potential for the town's economic recovery.[32]

Finding ways to finance the recovery of mining communities requires appropriate planning mechanisms at the time of project initiation. Companies usually need to have regulatory oversight on this matter to also avoid sudden closures or inadequate bonds for cleanup. In developed countries the rapidity of closure due to capricious commodity prices or political decisions can have an enormously negative impact on communities. For example, the Vatakoula Gold Mine on the Pacific island of Fiji operated for more than seventy years in close proximity to the fabled Fiji water facility. Due to political decisions by the parent company in Australia, operations suddenly ceased in January 2007, with not even a week's notice to the seven hundred workers at the mine. Located in a relatively remote part of the country, the community had no employment alternatives and was left to come up with new livelihood options entirely on its own.[33]

Closure of extractive enterprises should not come as a surprise, but the speed and efficacy of the process must be carefully calibrated to community needs. In their anthology titled *Coping with Closure,* Cecily Neil and associates state that "mine closure should not be looked on as a problem, but as the natural conclusion of the process of exploiting a finite resource."[34] Yet this is only possible if the resources to recover the social and physical landscape are adequately provided for at the outset.

The Evolving Balance

The topography of the earth is perpetually changing with the forces of plate tectonics, solar cycles, and celestial events such as a meteor strike. Mountains are eroded with time, craters form with volcanic eruptions, and species become naturally extinct with cataclysmic phenomena. However, human-induced changes to the landscape can be controlled to some degree and need to be considered with care. While we must dispense with an ossified view of the landscape, the functionality of natural environments as life-giving crucibles must also be conserved in any extractive enterprise. Even if the landscape is physically transformed by our intervention, the biogeochemical factors that may be invisible to the wondering observer need to be sustained. Livelihoods of future generations will depend on these ecosystems, as will the survival of the multitude of organisms with whom we share this marvelous planet.

An entire subfield of ecology has developed around the concept of "restoration," a term that is still contested at multiple levels.[35] Critics argue that restoration can only create replicas of natural systems, since human ingenuity cannot reach the level of complexity required to follow natural processes. However, the goal of a productive restoration effort is not to recreate the natural environment but rather to put in place structures and healing mechanisms that allow for natural processes to restore the landscape. Human restoration techniques are thus aimed at merely catalyzing the process of regeneration that is inherently present in natural systems. Restoration ecologists often speak of the "structure" and function of ecosystems—the former referring to the biodiversity of the environment and the latter to the chemical basis for the system's productivity in terms of nutrient availability and chemical balance. In order to achieve this task, there are some key principles to consider for planners of extractive enterprises in particular.

Watersheds are the key unit of analysis for a productive restoration enterprise. Rivers and their tributaries are the arteries of the landscape and provide the essential ingredient for all life forms to develop and regenerate after disturbance. The quality of water is thus the most significant litmus test for any ecosystem that needs to be regenerated. Extractive enterprises usually change water quality most acutely because they mobilize certain natural chemicals from the adjoining rocks or add contaminants that are used in the processing of minerals. All restoration efforts should therefore focus on the quality of water systems above and below ground before all other efforts are undertaken. Concerns about water quality need to be addressed at the root source rather than through symptomatic treatment of water. All too often, regulatory authorities are more concerned with fixing drinking water concerns by treating the water before it goes to consumer taps rather than improving the inflowing water in the first place.

Applying engineering principles to restoration is also fairly controversial, but, in the case of extractive enterprises, the landscape has been so severely altered that some level of structural engineering intervention may be essential to ensure functionality. Stabilizing slopes of mining landscapes, building channels for water drainage, and perhaps even purposely refrigerating the groundwater may be essential to ensure environmental functionality. Such interventions require community support to be successful in the long run, and a major educational effort alongside a comprehensive community engagement process is essential for success.[36]

All too often, "experts" are presented in restoration projects, but without community acceptance of the plans, projects can falter and fail. In most cases, communities are first concerned about their personal safety, followed by the potential for sustainable livelihoods after the restoration effort has been completed. In many cases of extractive industry, the restoration project itself can be a protracted multiyear affair that can create jobs on its own. However, these jobs are still inherently transient, and the apprenticeship that might come from such short-term enterprises needs to be considered as part of the employment recovery of the region.

All good restoration projects need to have a system of monitoring over extended timescales. There are few economic incentives to create such a monitoring system, however, and regulatory oversight is required to ensure that environmental criteria are studied on a regular basis. Technology has made such tasks much easier through remote testing systems for groundwater. The data collected still need to be studied and analyzed by experts who do not have a conflict of interest regarding the outcome.

Connections with academic institutions are extremely useful in such cases because students can play a vital role in data analysis from reclamation projects.

The overall impact of an extractive project will be felt for decades and centuries, perhaps even for millennia in the case of large open-pit operations. Even so, eventually the landscape will homogenize the impacts through natural processes. The endurance of life is astonishing in even the most extenuating circumstances. However, we cannot take the prospect of extinction for granted. Planning for restoration of degraded landscapes from the extractive sector will be a monumental challenge for generations to come, but it is an opportunity for ecological redemption that we should relish with care and creativity.

9

EMBRACING THE TREASURE IMPULSE:
TOWARD CAUTIONARY CREATIVITY

> The study of the substances of the earth's crust, of the air over and the waters under earth, which has led to our present knowledge has been more adventurous than many a great journey.... Into the unknown world of things upon the "sea that ends not till the world's end" scientists ventured, and came back laden with treasure greater than all the gold and precious stones ever taken from the earth.
>
> —Robert E. Rose, *The Foundations of Chemical Industry,* 1924

On the shores of North America's Lake Superior is a settlement called Twin Harbors, where in 1902 a remarkable group of five entrepreneurs launched a business venture to extract a mineral known as corundum. This is essentially the same mineral that constitutes precious rubies and sapphires, but Henry Bryan, Herman W. Cable, John Dwan, William A. McGonagle, and Dr. J. Danley Budd had little interest in gem-quality corundum. What motivated them was not the rare aesthetic appeal of this mineral but rather its physical properties of strength and hardness. Grinding stones and sandpaper were the products they had in mind when they sent out their first shipment of rock to be analyzed and sold. They didn't get much for their quarried rock, because it ended up having very little corundum and was instead dominated by a rather worthless mineral called anorthosite. Yet they were determined to find a solution to this dead end in their enterprise. With the help of a local venture capitalist, Lucius Ordway, the founding quintet was able to sustain the company, which they had named Minnesota Mining and Manufacturing.

Because of its rocky start in every sense of the word, the company was always striving to find new sources of materials and an innovative array of products. It eventually outgrew its origins as a mining company, and the name was changed to the 3M Corporation. The mining company reinvented itself as a leader in chemical innovation, from low-weight efficient insulation to the handy-dandy Post-It labels. What also differentiated the company was its willingness to procure ideas from its own employees, even if these might make some of its own existing products obsolete. If Post-Its were more efficient and less material-intensive, it was fine with the company if they competed positively with other pioneering products like its Scotch cellophane tape.[1]

More than a century after its inception, the company boasts over seventy thousand products that have emerged from its commitment to chemical creativity, spread over two hundred countries. Although the company has faced numerous environmental challenges, such as the contamination of soils near its facility in Cottage Grove, Mississippi, the corporate culture has remained committed to finding new and more efficient paths toward product development alongside environmental remediation. As early as 1975, the firm started an employee-based program of environmental innovation called 3P—a fitting abbreviation for Pollution Prevention Pays, which has kept 2.6 billion pounds of pollutants from entering the environment and saved more than a billion dollars worldwide.[2] It has been willing to spur innovation through environmental constraints, sending its scientists in the quest for chemical treasures that may be a bit more elusive to find in the short run but more promising in the long run. An interesting example of this was the company's decision to develop an alternative to volatile organic compounds (VOCs), which are extremely effective adhesives but have serious ecological and health impacts. It took three years longer to get "super-sticky" Post-It notes on the market, leading to an initial loss of $10 million in revenues, but the company found a better chemical in the process and the product is more sustainable as well.[3]

One company can't be a universal role model for transforming a material economy, but a story of such success can be emulated with relevant permutations. Many efforts will fail, perhaps be labeled as greenwash, may even be genuinely misguided. However, our contemporary environmental predicament can be resolved only by embracing our relations with materials, minerals, and the essential chemicals they comprise. The wealth generated by our attachment to materials in various forms has fueled economies in some cases but faltered on other accounts. Those who have benefited most from the extractive industries have to shoulder a tremendous responsibility for society at large. Embracing the treasure impulse will inevitably require us to consider how the accumulation of wealth in this regard can benefit society at large.

The Titans of Treasure

In early 2005, the largest single donation to an academic institution in history was announced. To the surprise of many, the recipient was not an elite institution in the United States, nor was the money from Bill Gates or Warren Buffett. Rather, the gift, in the staggering amount of $1 billion, was made by the mining tycoon Anil Agarwal to establish Vedanta University in the impoverished state of Orissa in his native India. Agarwal grew up in relative poverty in the squalid city of Patna in the state of Bihar and started out with a minor copper cable operation. He had remarkable business acumen but very little conventional education and faltered with even basic English. Yet he had the ambition to integrate his business vertically so that the primary resources used in his products would be under his control. Copper was a capricious commodity, with prices fluctuating wildly for much of the twentieth century. Finding the right opportunity, Agarwal bought a series of mines in India and Africa, setting up Vedanta Resources just as India was going through a privatizing phase. The timing was right for Vedanta to flourish, and Agarwal became one of the hundred wealthiest people in the world by the turn of the millennium. His company is the largest industrial taxpayer in India and has begun to diversify by bidding for service-sector businesses, such as privatized airports.[4]

Wealth has been accumulated in other sectors as well, such as computing. But for the developing world, some of the most striking rags-to-riches successes and consequent philanthropy come from the mineral sector. The story is by no means new, either. If we go down the list of some of the great philanthropists in history who made a global impact in their investment, we see that the origin of the wealth is often tied in some way to minerals. Georgius Agricola, the famed German father of modern metallurgy, observed as early as the sixteenth century that "of all ways whereby wealth is acquired by good and honest means, none is more advantageous than mining."[5]

While wandering through the halls of the building that houses the International Court of Justice in the Hague, Netherlands, one would probably assume that this great institution was built with donations from its member countries. Yet if one reads the fading plaques in some of the rooms or the fine print in the literature, it turns out that the benefactor of this "peace palace" was the Scottish-American steel magnate Andrew Carnegie, who donated $1.3 million in 1903 for its construction. Carnegie was a self-made industrialist who was especially concerned about the inequality that existed in societies, particularly between the laborers who toiled in mines and those who benefited from their ardor. In one memo to himself he wrote:

> There is no class so pitiably wretched as that which possesses money and nothing else. Money can only be the useful drudge of things immeasurably higher than itself. Exalted beyond this, as it sometimes is, it remains Caliban still and still plays the beast. My aspirations take a higher flight. Mine be it to have contributed to the enlightenment and the joys of the mind, to the things of the spirit, to all that tends to bring into the lives of the toilers of Pittsburgh sweetness and light. I hold this the noblest possible use of wealth.[6]

The size of the fortunes Carnegie and his philanthropic rival John D. Rockefeller accumulated needs to be seen in the context of contemporary holders of wealth and power to be appreciated (Table 1). Current billionaires do not even come close to the wealth generated by the tycoons of the industrialization era. Sam Walton, the founder of Wal-Mart, who is most commonly associated with material capitalism and mass consumerism, ranks seventeenth on the list of historical fortunes. The fortunes that Carnegie and Rockefeller built at the pinnacle of industrialization seem likely to remain unmatched.[7] Similarly, the power they wielded with their wealth has certainly not been matched since then.

To their credit, those who benefited from the production of minerals, particularly oil, often put their collected treasure into public service in

TABLE 1. The twenty wealthiest individuals ever

	Age at peak of earnings	Wealth in billions (in 2008 U.S. dollars)	Origin	Company or source of wealth
John D. Rockefeller	74	318.3	United States	Standard Oil
Andrew Carnegie	68	298.3	Scotland	Carnegie Steel Company
Nicholas II of Russia	49	253.5	Russia	House of Romanov
William Henry Vanderbilt	64	231.6	United States	Chicago, Burlington, and Quincy Railroad
Osman Ali Khan, Asaf Jah VII	50	210.8	Hyderabad	Monarchy
Andrew W. Mellon	80	188.8	United States	Gulf Oil
Henry Ford	57	188.1	United States	Ford Motor Company
Marcus Licinius Crassus	62	169.8	Roman Republic	Roman Senate
Basil II	67	169.4	Byzantine Empire	Monarchy
Cornelius Vanderbilt	82	167.4	United States	New York and Harlem Railroad, shipping
Alanus Rufus	49	166.9	England	Investments
Amenophis III	50	155.2	Ancient Egypt	Pharaoh
William de Warenne, 1st Earl of Surrey	38	153.6	England	Earl of Surrey
William II of England	44	151.7	England	Monarchy
Elizabeth I	69	142.9	England	House of Tudor
John D. Rockefeller, Jr.	54	141.4	United States	Standard Oil

(continued)

TABLE 1 (*continued*)

	Age at peak of earnings	Wealth in billions (in 2008 U.S. dollars)	Origin	Company or source of wealth
Sam Walton	74	128.0	United States	Wal-Mart
John Jacob Astor	84	115.0	Germany	American Fur Company
Odo of Bayeux	61	110.2	England	Monarchy
Bill Gates	44	101.0	United States	Microsoft

Source: *Forbes* Research Unit, 2008 (www.forbes.com/lists). Amounts are adjusted to current dollars.

various ways. Minerals centralized wealth with alacrity, and many of the individuals who started with nothing and became rich felt morally obliged to share their fortune with those in need. The inequality that was created in many cases by sudden influxes of mineral wealth was perhaps too stark to merit a simple justification on aptitude differentials. Those who benefited from the serendipity of mineral rushes perhaps felt a greater compunction and propensity to give.

Clearly Rockefeller's accumulation of wealth was also attributable to the lack of regulatory enforcement at the time regarding monopoly enterprises. This was arguably due to the fact that oil was such a fantastically transportable commodity that it soon came to be considered a panacea for energy needs. Rockefeller was quite conscious of the level of power he wielded and set out to focus on how best to use it constructively, with particular attention on education.[8] The range of his philanthropy was vast, from establishing the Spelman College for African American Women in 1884 to setting up the China Medical Board in 1914. Like Bill Gates, he was also instrumental in global health campaigns; the vaccine against yellow fever was largely developed because of his funding in a Rockefeller Foundation laboratory.[9]

The philanthropy of mineral tycoons was not confined to the Americas. Even during the age of colonialism, South Africa's "Randlords" were very conscious of the inequality generated by their fortunes and tried to find some philanthropic outlets, most notable among them being Cecil Rhodes, the founder of De Beers Corporation.[10] Despite his ardent support of colonialism and overt claims of British superiority in all world affairs, the endowment of the Rhodes scholarship at Oxford University in his bequest is considered one the most significant acts of global educational philanthropy. Such notable politicians as Bill Clinton have benefited from this award, and the title of Rhodes scholar is still among the most coveted biographical pointers among circles of influence.

Every spring *Forbes* magazine releases a list of the world's billionaires, an exclusive club of around eleven hundred people. The list is eagerly awaited to see the ranking of celebrity tycoons like Bill Gates. Additionally, what matters to many is how these individuals made their fortunes. The list announced in spring 2008 recorded total net assets of just over $4.3 trillion. Clearly most of this money is still not destined to help the poor and stands in stark contrast to the world's total annual development assistance, which is on average only around $100 billion.

Philanthropy by individuals often turns out to be a rewarding yet risky social policy tool, given the idiosyncrasies of the elite. Many of the nouveau riche prefer to spend on themselves or support their own pet projects, such as buying artwork to show national pride. The Russian-Uzbek tycoon Alisher Usmanov made much of his fortune in extractive industries. Usmanov is spending several hundred million dollars buying Russian artwork in private collections and purchasing mansions, while his ethnic homeland languishes in abject poverty. Despite the roster of philanthropy by tycoons, studies of charitable giving repeatedly show that wealthy individuals still give a much smaller proportion of their money than the middle class or even the more impoverished.[11] So relying on individual giving as a social policy tool is not an efficient or viable option on its own. One of the greatest challenges of our times is to consider how to spread the wealth from our treasure impulse, which tends to asymmetrically benefit some at the expense of others.

Those who are willing to take risks are often rewarded disproportionately over those who don't. The propensity for taking risks clearly has some biological roots: this tendency is found in other advanced organisms as well, with disproportionate outcomes. A study of macaque monkeys in 2005 found that there was a tendency to follow a riskier target that might provide different amounts of juice over a safer one that always gave the same amount of reward.[12] It is no wonder gambling has been found in societies since time immemorial. Although our treasure impulse may structurally be predisposed to outcomes that will be unequal, we cannot use that as an excuse for leaving millions to languish in unfulfilled aspirations.

Nobel laureate Muhammad Yunus imagines a world of the future where poverty will be seen in museums and people wonder how we could have let our fellow humans be so deprived when there might have been enough treasure to triumph over destitution. In *Star Trek*, Gene Roddenberry went a step further and described a distant future where humanity has adequate governance structures that transcend financial and political constraints. The earth would be essentially devoid of scarcity, and our most prized medium of exchange, money, would be entirely absent from planetary life.[13] The only civilization to use money in this fictional creation are the primitive Ferengis, who are universally despised for their lack of material ethics. Whether such a universe will remain only a product of imagination remains to be seen. But what we clearly need to do in this moment is to seek paths of channeling our quest for material gain more constructively and find better ways of sharing our collective treasures.

Whether this is done through trust funds for communities, social safety nets, global development funds from mineral extraction revenues, or individual philanthropy, we cannot afford to lose sight of those who just couldn't find the booty in the treasure hunt. Regardless of whether aptitude or pure luck is the determining variable, finding a way to meet some globally agreed upon basic human needs is essential. Our treasure impulse may well be the only way to achieve this task, but it will need to be appropriately configured through ethical behavior, regulatory persuasion, and social enterprise.

Sustainable Polarity?

Sustainability for our purposes is the attribute of a process that can harness a resource while allowing for replenishment of its base capital (natural, economic, and social) to meet future needs. The future needs of a community must be defined in terms of technological forecasting as well as physical resource constraints. Considerations include social and community market economics. Social implementation of sustainable processes is important, but must not overlook the reality that many projects that are socially satisfying can be utterly unsustainable from a physical perspective. Technological innovation has performed wonders, but it cannot defy physical constraints. On the other hand, in the event of social resistance to a physically sustainable process, there is greater leverage to change social views through education, training, and economic means. Thus, the social concerns are important, but only after the physical sustainability of a project is fully understood. So the question remains: are extractive enterprises compatible with sustainable development and hence congruent with a long-term vision of security? The answer must consist of two parts.

First, there is no doubt that extractive industries under present technological conditions do have a certain degree of permanent impact on a region. Mineral industries also involve extraction of nonrenewable resources. By these measures, the answer at one level is: no, mining is not sustainable. However, while the landscape may be permanently changed by extractive enterprises in certain ways, that does not necessarily mean that communities cannot thrive if projects are appropriately planned. Extractive enterprises can therefore be a prelude to sustainable development if the community is willing to absorb a certain degree of permanent impact. The key, then, is to be able to use extractive industries as an entry point toward a more stable industrial or service-based economy that is not inherently obsolescent. Economic diversification is the operative phrase for both interstate and intrastate security. This applies to countries diversifying their energy sources, as well as their industrial investment. Indeed, one of the less compelling aspects of the arguments of resource-war determinists regarding U.S. security interests has been the ongoing diversification of U.S. oil and gas sources. Furthermore, the United States gets almost half its oil from domestic sources, and the largest foreign contributors are its immediate neighbors—Canada and Mexico. American energy policy is trying to diversify away from dependence on Middle Eastern oil, and thus the argument that the U.S. government is going to war in the region simply over oil sounds less convincing, especially given the high financial and human cost involved.

Another interesting rebuttal to resource-war determinism is that dependence can often lead to more cooperation than conflict, as observed with water resources, where hardly any wars over water have been documented in recent memory, despite much saber rattling.[14] That the Taliban visited Texas and met with executives of gas companies before September 11, 2001, could be considered a mark of how resource interests can potentially foster cooperation between two utterly different cultures, rather than fueling conspiracy theories about conflict.[15] Had a deal for technological cooperation on natural gas extraction been brokered during this visit, would potential chances of conflict been averted? The key to a fruitful outcome depends on how governments in extractive countries handle their resource endowments, and what international mechanisms exist to moderate the wealth and woe that they might bring. Education is a critical issue in many areas where windfall development is to occur, and the establishment of schools and other vocational training programs independent of their utility to the developers is critical. Hence a contingency development plan would be essential, particularly for projects of planned obsolescence, such as mining.

Nevertheless, alternative development strategies should be strongly encouraged for any risky windfall venture. This would likely require regulatory enforcement, because for a developer it is a classic externality. The specifics should remain flexible, however, given the highly diverse nature of alternatives that may exist for various communities. Some examples of successes and failures in this context need to be further studied with the aim of testing various hypotheses regarding the efficacy of extractive industries and their influence on intrastate and interstate security. These hypotheses are likely to focus on five areas: stakeholder engagement, including negotiation versus consultation and transparency of the process; supply-chain management of the mineral resource; pollution prevention and risk management systems; post-closure remediation and sustainable livelihoods; and potentially cooperative linkages of projects through mutual dependence. Clearly some of the process issues, stakeholder involvement in decision making, organizational dynamics, and implementation would need to be ironed out for such a system to be effective. But extractive industry projects must not take place in a vacuum, and they should be catalysts rather than reactants in the synthesis of development.

It is true that extractive industries have become far safer and more responsible than the indentured mines of yesteryear, but there are still continuing concerns about not only the extraction but the responsible use of the minerals. In January 2006, the coal-fired Mojave power station in Arizona was closed for noncompliance with air-quality standards. This station provided power to 1.5 million homes in Arizona, Nevada, and California, and was fueled by coal from the Black Mesa and Kayenta mines on Navajo and Hopi lands. Because of the plant closure, the mines have had to close as well, leading to enormous economic pressure on the tribes. Unlike the community consensus against uranium mining, coal mining had been the major source of revenue for both tribes.

For the past three decades, these communities have struggled with the pros and cons of energy from coal and the water depletion that resulted from having to convert the coal into a slurry for transport to the power station more than 250 miles away. While the mine provided over $85 million in annual benefits to the Navajo and around $10 million to the Hopi, there had been repeated concerns about how best to manage the resource. The Hopi had recently disallowed further water pumping from one the aquifers and had negotiated to get water from a larger, more resilient aquifer. Now environmental violations have led to the demise of the mine and a loss of livelihoods for many Native Americans. However, this may open the way for other forms of creative enterprise that can perhaps build a post-mining economy for the community. At least through effective planning and management of their mineral royalties, the Navajo have accumulated several hundred million dollars in a trust fund.

Sustainability with reference to the extractive industries is a game of lifestyle choices versus innovation under constraints of time. If all goes well, the pace of technological innovation and substitution will keep up with extraction and usage. If there is some lag between innovation and depletion, however, we will have to contend with a series of crises. To improve the chances of avoiding this problem, we need to consider a full range of efforts, including extraction in new areas, conservation and efficient use of current supplies, recycling, recovery of embedded materials, and the search for renewable substitutes where possible. Cornucopian scholars such as Julian Simon were perhaps too sanguine about our speed to innovate, whereas Cassandran researchers like Paul Ehrlich are perhaps too pessimistic about human ingenuity.[16]

The way to possibly succeed in this pursuit of sustainable resource use is to move away from either of these polarities and consider the approach from a pragmatic perspective of harnessing technological innovation within ecological constraints. We have been wrong in the past about our mineral projections, particularly of nonfuel minerals, and we need to spend more effort in understanding why we were wrong and how to refine our estimates.[17] At the same time we need to consider our material usage with a clear eye toward sustainability. Veteran materials-flow researchers Thomas Graedel and Robert Klee have urged us to "get serious about sustainability" by considering four areas of policy intervention: establishing the available supply of the chosen resource; allocating the annual permissible supply according to a reasonable formula of market process; establishing the "recaptureable" resource base; and deriving the sustainable limiting rate of use and comparing that to the current use to calibrate regulatory intervention.[18]

In the meantime, the quest for some agreeable permutation of molecules from geological deposits, biological sources, or chemical synthesis must also continue. Are there any win-win prospects for a material and energy panacea? Not quite, but there are some very promising features to consider. Much of our mineral exploration has thus far been on land, with the exception of oil and diamond extraction that have moved offshore. The potential for deeper-water drilling of mineral deposits might well prove to be a win-win solution that can buy us much more time from the clock of nonrenewability. Investors are flocking to companies, such as Nautilus, that are actively engaged in deep-water exploration for minerals. International law has also recognized this prospect under the Law of the Sea Convention, and an International Seabed Authority has been established to consider mineral claims for the deep. In the 1970s, there was a strong interest in mining manganese, and over $700 million was invested, but the technology was not quite ready for the undertaking. However, it appears that technology has caught up with this task, and even the larger mining companies are investing in aquatic mining efforts, particularly off the shore of Papua New Guinea.[19] Environmentalists such as Sylvia Earle have bet on offshore technologies as a win-win outcome for future resource extraction. She is confident that many resource extraction efforts can be moved offshore with minimal environmental health impacts.

Many other transitional technologies might yet be needed before we can find a more lasting source of energy and material extraction. Would algae provide abundant renewable energy as liquid fuel as well as sequestering carbon? Perhaps the production of methane from biotic sources will be the next frontier in energy generation. Much remains to be seen in these two particularly promising avenues of material inquiry. Unless we are victims of some cataclysmic cosmic event, human societies will continue to seek out new ways of harnessing ideas out of our material world.

Industries dealing in materials used primarily for luxury goods are desperately trying to find more "need-based" uses for them as well, as a means of justifying greater use. This would provide greater legitimacy to mining efforts as well. Only 10 percent of gold mined in 2007 was used for industrial purposes, such as in electronics and dentistry; 60 percent was made into jewelry and the rest was used for investment purposes. This may well be a good sign of how our wants might spur need-based innovation. The gold sector is seeking new uses for the precious metal in air conditioners, firefighters' masks, and anti-cancer drugs. Researchers have found that certain gold-based compounds can selectively target the mitochondria, which are the energy-generating system of a cell and which vary significantly between cancerous and healthy cells. This is different from many current chemotherapy drugs, which target the cellular DNA, and can end up also damaging healthy cells. A research project funded by South African gold producers is close to commercializing a gold catalyst that cleans air of carbon monoxide. In the words of Elma van der Lingen, the leader of Project Autek, "We're trying to get people to change their perceptions and show that gold is not just there for its beauty, but that it can be useful as well."[20]

Scenarios of Fortune

The completion of two thousand years in the Gregorian calendar was a time of much ambition and anxiety all over the world. The United Nations initiated the Millennium Development Goals, with clear targets to reduce global poverty in the following twenty years. Various governments funded the work of ecologists conducting a Millennium Ecosystem Assessment, to give us a planetary health check. Computer scientists fretted over the turning of the zeros and the dreaded 2 making its way into that thousands-digit space that had been occupied by a 1 for a thousand years.

Beyond all the hype about Y2K and apocalyptic predictions of technological paralysis, the world benefited from this time of concerted reflection. We paused to think about the passage of time at a civilizational scale. Our ancestors a thousand years earlier never planned for our predicament in 2000, but we were able to at least consider planning for the following thousand years.

In the millennium mood, the mining industry also considered ways of reflecting on its past performance and its contribution to sustainable development, which was very much in vogue across all businesses. The timing was right with all the momentum toward global self-assessments, and the growth in research being undertaken before the World Summit on Sustainable Development in 2002. Ten of the world's leading mining companies joined forces to fund the Mining, Minerals, and Sustainable Development (MMSD) initiative in 1999. The work of the MMSD would involve high levels of expert research and participatory consultation to generate a report considering how a nonrenewable sector such as mining could still be part of a larger equation of sustainable development. The central premise of the effort was to consider material flows and capital flows to catalyze infrastructure growth and pave the way for other renewable industrial sectors to also take root in extractive economies.

Spread out across five continents, the MMSD initiative became a bonanza for environmental management consultants and itinerant academics. The North American arm of the initiative was coordinated by the International Institute for Sustainable Development in Manitoba, Canada, which decided to employ "scenario analysis" to consider the future impact of minerals. Based on military strategic planning exercises following World War II, the exercise involved an intensive three-day workshop in which various stakeholders from academia, industry, civil society groups, and government were asked to consider variables that could affect the development trajectory of mineral economies. As a participant in this process, I came with initial skepticism that this might end up being another vacuous exercise in unraveling the Gordian knot of sustainability. However, the level of conversation proved to be fairly detailed and robust as specific examples of mining closure plans were discussed with engineers, planners, and economists. The result was four scenarios centered on a Cartesian plane with two axes indicating the trade-offs between economic performance of the sector and societal values (fig. 20).

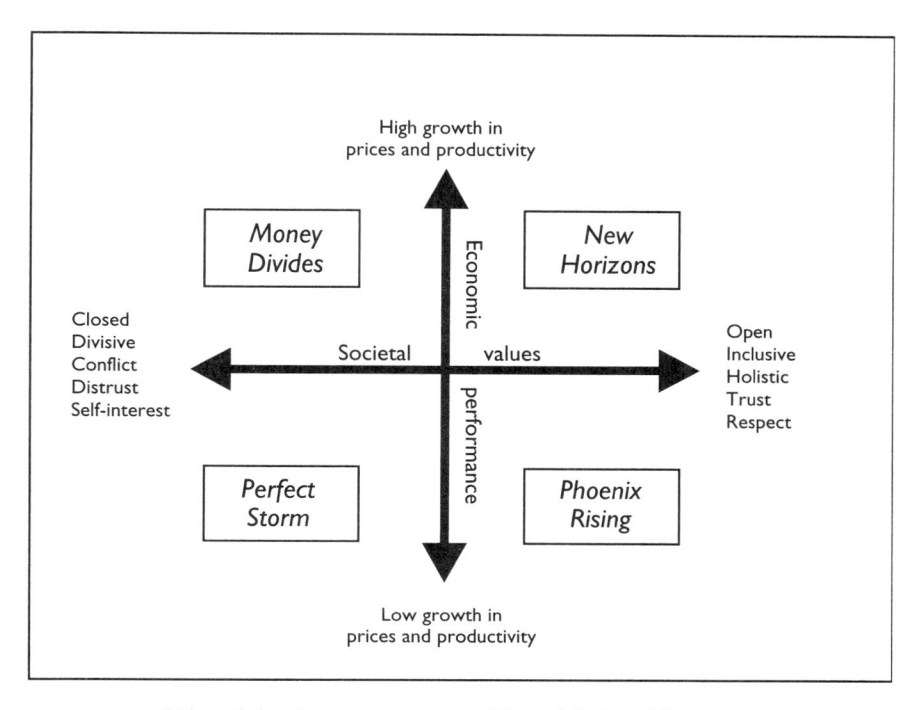

```
                    High growth in
                  prices and productivity
                           ↑
        ┌─────────┐       │        ┌─────────┐
        │  Money  │    Economic    │   New   │
        │ Divides │       │        │ Horizons│
        └─────────┘       │        └─────────┘
Closed                    │                    Open
Divisive                  │                    Inclusive
Conflict  ←── Societal ───┼─── values ──→      Holistic
Distrust                  │                    Trust
Self-interest             │                    Respect
                      performance
        ┌─────────┐       │        ┌─────────┐
        │ Perfect │       │        │ Phoenix │
        │  Storm  │       │        │ Rising  │
        └─────────┘       │        └─────────┘
                           ↓
                    Low growth in
                  prices and productivity
```

FIGURE 20. Mineral development scenarios (From Mining, Minerals, and Sustainable Development [MMSD] North America, *Learning from the Future* [Winnipeg, Manitoba: International Institute for Sustainable Development, 2002])

The quadrants represent a combination of outcomes of these two critical variables. For example, the upper right quadrant characterizes a future in which societal values become more integrated and inclusive, fostering an environment of respect and mutual trust. At the same time, a healthy economic environment is predicted in this scenario with robust commodity prices, growth, and productivity improvement in the minerals industry. This is the win-win outcome that might be possible if all the intervening variables are appropriately aligned. The exercise considers what pathways might lead to such a scenario in detail, with deliberative interactions between all stakeholders involved to see what would be the behavioral responses to each step of the process. One can then work back from the scenario to see what factors are within our control in terms of planning and which are unpredictable and need to be considered in any risk-analysis calculations.

Here are some of the key factors considered in the development of each scenario.[21]

1. New Horizons: The mineral industries increase their investment in sustainable development. In large companies, social and environmental issues receive an enhanced profile internally, as reflected in budget allocations, training, and a variety of programs. This same commitment is found in a majority of small companies. Industry associations take a significant leadership role in translating sustainable development concepts to a scale that can be put into practice by smaller companies, from exploration to intermediate operations. At the same time, recycling of materials increases considerably, and erstwhile mining companies begin to see themselves as material service providers rather than extractive industries per se. Oil companies reinvent themselves substantially as energy service providers, while conservation efforts are able to keep pace with technological advance in finding alternative energy sources.

2. Phoenix Rising: Social pressures initially push the mining and minerals industry to the limits of survival. Low economic growth undermines demand and real prices decline for most metals and minerals. At the same time, public environmental and social concerns remain high, and nongovernmental organizations are influential in raising their concerns. As time passes, the determination to change grows. Innovative practices are initiated. Dialogue is opened with local communities and a variety of nongovernmental organizations and other interest groups. New managers with new attitudes attempt to bridge the gap of distrust. By 2015, the result is a transformed mining industry. Responsible practices, innovative procedures, and new technology have lowered costs, increased efficiency, and expanded opportunities for new mines. In the United States, steps are being taken to redraft the general mining law of 1872.[22]

3. Perfect Storm: Mining in the early years of the new millennium continues to struggle as commodity prices begin to fall from oversupply, similar to what plagued the industry in the last two decades of the twentieth century. Low economic growth creates insufficient demand for metals and minerals in the former boom regions of China and India. Investor interest in mining shares continues to languish, and mining share values

remain depressed. New oil and gas sources are not found as easily as expected, and energy costs become exorbitant. Investment firms lay off their mining analysts, and equity capital markets are effectively closed to the industry. Environmental damage also diminishes the risk tolerance of society to mineral extraction. In the meantime, few alternatives can be found to replace minerals, and there is a worldwide economic depression. Human societies have no choice but to curtail their consumption patterns drastically. Airplane travel and personal ownership of cars become luxuries for the elite—"a long emergency ensues."[23]

4. Money Divides: Expanding economic conditions stimulate prices and the financial prosperity of the mining industry in North America. Share prices rise, and there is a new sense of conceited confidence among executives in the industry, which leads to complacency. Mergers and acquisitions reflect increasing cash flows and a general sense that money can surmount almost any challenge to the sector. There is renewed investment in exploration, research and development, and automation technology. Labor demands increase. Some involve strikes, but these are resolved with wage increases. Executive bonuses expand, and the industry's leadership revels in its renewed wealth and optimistic outlook. While the economy and economic wealth are growing, national and personal security concerns are high. Income disparities are rising. Anti-globalization groups are becoming stronger and more sophisticated. Attention is increasingly focused on the abusive power—real or perceived—of large multinational corporations. This paradox of struggle in the face of prosperity persists for some time, but it is not sustainable. A cyclical downturn in markets is inevitable. A future of rising conflict and shrinking margins looms on the horizon, and a perfect storm seems to be on the horizon.

Ascertaining the likelihood of any of these scenarios, or permutations of them, at various geographic scales is a daunting task. Computer models and geological measurements may assist in the process of imbuing such scenarios with valuable data. Many of the projections from such ventures may end up being wrong. However, we must constantly strive to refine our ability to understand our material constraints as we consider the ecological challenges that lie ahead for us.

At the micro level, each extractive project that is to be initiated has to consider some specific questions to improve the chances of getting to a "new-horizons" scenario. Participants in the MMSD initiative suggested that there are seven questions that need to be clearly addressed before any mineral development project should be embraced:

1. Engagement: Are communities part of the negotiation process for any development effort?
2. People: Will the well-being of inhabitants in the area be maintained or improved, as measured by objective welfare criteria?
3. Environment: Is the integrity of the environment ensured over the long term?
4. Economic viability: Is the project or operation economically viable, and will the community be better off as a result of the project in the long term?
5. Traditional and nonmarket activities: Are livelihood potentials, such as subsistence hunting, going to be adequately protected during the course of the project?
6. Institutional arrangements and governance: Are rules, incentives, programs, and capacities in place to address project or operational consequences?
7. Synthesis and continuous learning: Does a comprehensive analysis show the net outcome of the project to be positive, and how will periodic assessment validate this finding through the course of the project?

One year after the MMSD report was presented to the delegates at the World Summit on Sustainable Development in Johannesburg, South Africa, these seven questions were used empirically in a deliberative process. The chief of the Tahltan/Nahanni Band of Athabascan indigenous residents of northern British Columbia, who had participated in the process, decided to use this structure to consider the prospect of mining with his community. Over a four-day symposium the community members considered their options and in what shape and form mining would be acceptable to them.[24] They decided that, with appropriate regulatory safeguards, mineral development could take place, but could not be rushed through without free prior and informed consent of the community.[25]

Empowering communities where mining can occur with veto power over a project is a highly contentious issue, since companies claim that mineral deposits are geologically determined, so they can't simply pack up and site a facility elsewhere (as is often the case with manufacturing plants). Community veto over projects can also be a problem for planners or scenario developers, and may also be economically inefficient. This is where we may want to resurrect some of the earlier tensions between incremental decision-making and planning ideals that are predicated in the concept of comprehensive rationality.[26] Is there some value on ethical grounds, as well as the quality of the output, to allow for a "science of muddling through" in decision making? Indeed, in a liberal democratic society, all planning efforts must be considered with some trepidation and remain flexible to changes in community attitudes, in the same way that some level of choice in consumer preferences spurs creativity. The key boundary in the spectrum of individual choice in decision making should be ecological constraints. Within such constraints, we can often spur even greater creativity and perhaps move toward a scenario reminiscent of "phoenix rising."

Yet there is tremendous fear among environmentalists about the slippery slope of creative enterprise that is so often championed by economists. At the heart of the controversy is the tenacious allegiance to economic growth among conventional economists and the equally adamant antipathy to growth among ecological economists. Creativity is often linked to economic growth in the minds of many analysts. In contrast, "steady-state" economics, a term coined by Herman Daly at the University of Maryland, has become a common refrain for the ecological economist.[27] However, economic growth has empirically been an important driver in development and has led some economists, such as Benjamin Friedman, to argue for its moral predication. Indeed, there is not a single example of a country reaching modern indicators of human development in the absence of economic growth.[28] But once that threshold of development is reached, further growth as measured by conventional indicators might lead us into a twilight zone of unsustainable behavior.

Unfortunately, where models of growth have been questioned or criticized by classically trained economists, such as William Easterly, environmental factors have largely been neglected. Instead, a colonial narrative has been invoked that can only escalate further defensive posturing among the dominant strains of economic thought.[29] The globalization discourse has also confused thinking in this regard by resurrecting misplaced visions of emancipation (as exemplified by the World Social Forum), on one hand, and institutional critiques of development donors on the other. Such well-intentioned activists and visionaries as Bill McKibben have also conflated some of the virtues of community capital and used it to advocate a rather insular vision of a "deep economy."[30] Although the role of community networks in reducing ecological harm is important at one level, the larger issues of poverty eradication must be considered at a global level if we are to have a pragmatic inculcation of environmental norms.

Must inequality be endured as a necessary storm surge until the rising tide of economic growth lifts all vessels to safety? This was the prognosis in the famed Kuznets curve, which suggested that inequality would increase with economic growth but eventually decline.[31] The same logic was also employed by economists to environmental harm, suggesting that ecological damage was a price to pay for initial development, after which a self-correcting mechanism would somehow kick in to improve environmental performance. However, the empirical evidence has only marginally supported the reduction of inequality and environmental harm.[32] Indeed, much of the research is now suggesting that we need to pay far more attention to indicators of environmental harm, such as ecological carrying capacity, to prevent irreversible damage to particular ecosystems that also sustain livelihoods. One Chinese government study in 2006, cited by the New China Agency, suggested that the country's western provinces will suffer an annual loss equivalent to $20 billion, or 13 percent of the region's gross domestic product, because of environmental damage.[33]

Are there better ways to reconcile inequality, livelihoods, and the environment than conventional indicators of income or wealth may lead us to consider? Perhaps the most promising arena in this regard is research originating from the Santa Fe Institute under the leadership of Samuel Bowles, which has provided a refreshing eclecticism to the debate.[34] The research suggests that globalization challenges egalitarianism, but that if we want to spread the wealth from rich to poor countries, ultimately some form of global trade is essential. Unfortunately, any mention of egalitarianism still conjures images of communism in the minds of many economists, and the Marxist stigma is quickly placed on such scholarship.[35]

The concerns raised about inequality by such research are also very much in line with the kind of incentive-driven policy that is the hallmark of capitalism. Consider, for example, the current pay scale for executives of corporations—the average is usually in the tens of millions of dollars, and may be hundreds of times that of other employees. What incentives do such executives have to either perform well or spur creativity, when they know that even after being fired they will have a lifestyle far superior to that of more than 99 percent of their employees? The incentive structure is skewed against positive performance and also against a constructive channeling of our treasure impulse. Some level of differential outcomes is important as a mark of healthy competition spurring growth and innovation. However, when the differentials become so acute that the competitive structure of the process is itself threatened by individual wealth, then we have a problem. The question, then, is what differentials of remuneration are allowable? This can clearly be determined through a deliberative process across wealth-generating entities—beyond a certain point of remuneration, innovative programs can channel additional bonuses into charitable trusts to be managed by the executive who earned that wealth. Clearly, some global regime would be needed in this regard to avoid capital flight and offshore accounts. Utopian as such a system might seem today, addressing such matters would not only be essential for sustaining human security but also would likely be most economically efficient.

The World Bank has been the stage on which many of the dramas about inequality, trade, and fair play have been played out at the international level, in how incentives are configured for the economic performance of countries. Vitriolic exchanges between celebrity economists like Joseph Stiglitz and Kenneth Rogoff have taken place in this context.[36] Several disgruntled World Bank employees, including Stiglitz, Easterly, and Herman Daly, have managed to claim legitimacy in their subsequent public critiques of the bank. Equally strident defenses of World Bank leadership have also come forth from such writers as Sebastian Mallaby.[37] At the core of these exchanges is an inability of either side to address issues of massive inequality, which is seminal in our struggle for charting development paths, while also considering our material dependence.

Consider a survey of ten thousand Africans from ten countries across the continent conducted by Globescan in 2007. The fundamental question asked was: What should be our government's top priority? The results reflect the salience of livelihoods to residents of the world's most

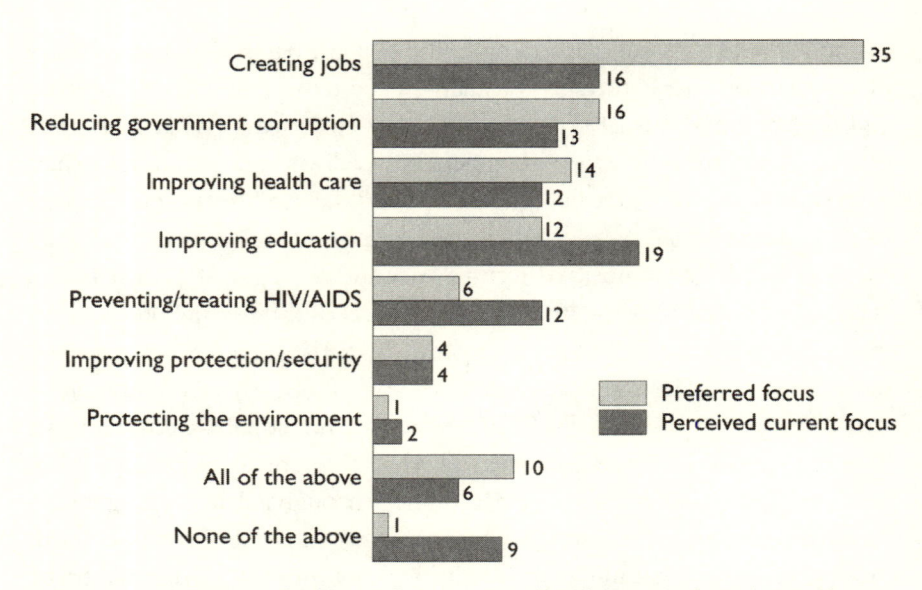

FIGURE 21. Priorities of Africans, from a survey by Globescan of ten thousand people in ten countries. Figures shown are percentages of all responses. (Commission for Africa, 2007)

Bar chart data (Preferred focus / Perceived current focus):
- Creating jobs: 35 / 16
- Reducing government corruption: 16 / 13
- Improving health care: 14 / 12
- Improving education: 12 / 19
- Preventing/treating HIV/AIDS: 6 / 12
- Improving protection/security: 4 / 4
- Protecting the environment: 1 / 2
- All of the above: 10 / 6
- None of the above: 1 / 9

impoverished continent (fig. 21). Finding jobs may well be the most significant issue for Africans, whereas protecting the environment is of least priority. But the reality is that both may well be inextricably linked. Also, making a linear argument for any industry simply on the basis of jobs can be problematic, if the jobs being created are harmful to society—for example, the huge employment created by the arms trade or drug trafficking. Instead, what is needed is a consideration of opportunity costs of particular forms of employment with a view of livelihoods that considers various potential paths to development in sequence. Efforts should be made to foster livelihood trajectories that are most viable in the long term.

To conserve our opportunities for either natural regeneration of renewable resources or the technological advancements to find substitutes, some scholars have suggested that we slow down the production of goods.[38] Although more modest consumption patterns can certainly help in stretching our time frame toward sustainability, if this strategy translates into "slowing down," industrious human behavior might not solve our problems either. The drive and dynamism that have been a hallmark of our impulse for seeking treasure must be sustained with appropriate incentives to focus on industries that can find creative ways to meet societal needs that are inherently predicated in materials. Reducing the pace of progress is not the answer to our woes; rather, finding effective ways of managing progress with the societal good in mind is. This change will inherently require more of our university graduates to consider environmental careers that are based on a better understanding of materials and earth sciences. Nudging humanity toward such socially beneficial avenues of endeavor may be the appropriate role of government.[39] Making an effective linkage between jobs and the environment, development planners will need to find ways of creating livelihoods with minimal ecological impact. Natural resources will inevitably play an essential role in this process, but we will need to channel our impulses for seeking such treasures in more constructive ways. Establishing a responsibly operated mine would be considered an opportunity, but so would finding ways of harnessing resources from aging infrastructure and employing traditional practices of energy conservation.

If there is a nefarious necessity in spurring and channeling our treasure impulse in an ecologically and economically efficient direction, it is perhaps the specter of regulation that tends to lead either side to entrenched positions. Corporate social responsibility in the environmental arena clearly has its limits, given the structure of markets. As noted by David Vogel in *The Market for Virtue,* "the main constraint for the market's ability to increase the supply of corporate virtue is the market itself."[40] If we are to have payments for ecosystem services that are conserved, for example, some form of regulation is likely to be essential to allocate costs and benefits for corporations. The common good of environmental protection will have a political price that pits proponents of individual liberty against the regulators. Yet in this day and age, many of the erstwhile proponents of individual choice on the conservative edge of the political spectrum are now willing to curtail individual liberties for goals of national security. Surely, the physical constraints of the environment's capacity to support and sustain human societies would merit due consideration on similar terms. It is no surprise then that vanguards of conservative doctrines, such as former speaker of the U.S. House of Representatives Newt Gingrich, are calling for a "Contract with the Earth."[41]

Nevertheless, even if regulation is liberally applied in matters of environmental security, there are still certain fundamental individual liberties that we have now come to accept as beyond the reach of regulation. Regulating birth decisions, for example, would have been the single most potent regulatory mechanism for the ardent neo-Malthusian in mitigating resource depletion and environmental harm, but is no longer plausible as a policy choice. China regulated population through denial of any services to families with more than one child, but this too has drawn rebuke from human rights activists.[42] Gone are the days when scholars like Garrett Hardin were proposing "lifeboat ethics" that advocated apathy toward the poor and the elderly or condoned a demise of populations to sustain "spaceship earth."[43] Environmentalism has been more universally humanized, but it has also led to a conundrum of how best to address our fundamental resource constraints. Even ardent proponents of population control have mellowed their conversations on the matter, considering the enormous ethical implications of such rash rhetoric.

The great economist and philosopher Amartya Sen recognized the challenge of reconciling efficiency and "optimal" societal behavior with liberalism several decades ago in his famous essay "The Impossibility of a Paretian Liberal."[44] Sen was concerned with human propensity for conflict when certain inalienable values collide within a liberal system that may also be trying to achieve Pareto optimality. Named after the nineteenth-century Italian economist Vilfredo Pareto, this condition could be theoretically achieved when a movement from one allocation to another can make at least one individual better off without making any other individual worse off. Sen showed that when we define certain unrestricted domains of human behavior, such as decisions on having children or what we wear or eat, then we cannot aspire to having an optimal society. Kenneth Arrow arrived at a similar insight with regard to voting behavior in his famous impossibility theorem, which suggested that voting systems are not capable of converting the ranked preferences of individuals into a community-wide ranking of societal preferences.[45] Our approach to consumption and the environment must grapple with this fundamental challenge, and therefore a multifaceted approach with incentive-driven regulations, technological innovation, and literacy-based behavioral change is essential.

The End of Everything

Humans have a constant and troublesome yearning for a beginning and an end to every story. We not only yearn for closure, but also try to project narratives that lure us toward that finality. Our own helplessness in the larger cosmic scheme may lead us to consider our fate with a sense of inevitability and futility. The human mind is structured around an intrinsic notion of start and finish, because most phenomena in our lives are such—starting from our birth and ending with our death. Even while relishing the remarkable accomplishments of science and technology that were unfathomable only decades earlier, we are still attracted to fatalism. For scientists, there are certainly limits that human ingenuity cannot exceed, such as the laws of physics and cosmology. Within those limits, however, the enterprise is still immense, reminiscent perhaps of that fixed line from geometry class with a finite distance but an infinite number of points along its length.[46]

Veteran science journalist John Horgan posited in 1996 that we were reaching the "end of science," suggesting that most of the important scientific discoveries that can be made have already happened and now we need to apply what we have to human progress. Our treasure impulse might get us only so far in our material quests and the permutations of molecules that can provide for our needs and wants. Horgan may be correct in that the parameters of science have been established and there is a certain irreversibility to knowledge acquisition because we are able to empirically observe phenomena at the atomic level. However, we are far from ending our quest within those parameters. Even the inevitability of aging is being challenged by the researcher Aubrey de Grey, who is investigating avenues to slow or reverse the aging process.[47] This would be a Malthusian nightmare for many environmentalists, but one they scarce oppose on most ethical principles, so long as the quality of life was acceptable to the individuals in question.

Other writers have pondered the end of nature, the end of faith, and, most memorably, the end of history. Although proximate examples may substantiate any of these arguments at some levels, the larger uncertainty about our planetary potential and our abilities as a species remains. Futurists such as Raymond Kurzweil have suggested that humanity is reaching a critical point in its civilizational trajectory—not an end, but rather a new synthesis of biological and mechanical invention.[48] He calls this transition a "singularity," because the implications will be perhaps as significant for humanity as the unique cosmological event that created the universe. Kurzweil's main premise is that, with the advent of nanotechnology and biological computing, we are moving into an entirely new era of our relationship with materials and the development of human civilization. His idea is particularly reassuring to those who wonder what would happen to "menial" professions if everyone were educated and aspiring to doctorates. Machines could then take over some of the more humdrum duties of life at a much larger scale than they already have.

Jerald Schnoor, the editor of *Environmental Science and Technology*, has, however, cautioned futurists like Kurzweil about the limits of technology as follows: "Americans believe that technology will always pull us through. From the Manhattan Project to the Green Revolution to putting a man on the moon, technology has always been the savior. And technology will certainly play a role in our transition from the fossil fuel age. But the human dimension of how to organize ourselves to save the planet is the most daunting challenge of all. The U.S. must provide leadership, compassion, and inspiration to reverse this problem. It will require a change of heart, not simply a change of mind."[49] This warning is well taken, and certainly all efforts to stimulate environmentally conscious behavior must be heeded. The insurance argument, reminiscent of Pascal's wager (to believe in God just in case it is true), makes sense for a risk-averse society. However, it must be done in ways that do not stifle our inquisitive and exploratory tendencies, which may be the most likely pathway to salvation of the species. No matter how much we conserve, we will ultimately be faced with resource constraints, given our ethical stances on population control and the sanctity of human life. In essence, this is the kind of environmental pragmatism that was envisaged by such scholars as John Dewey a century earlier.[50] This approach is notably different from the politically divisive rhetoric of environmental revisionists, such as Bjorn Lomborg, who focus entirely on decoupled prioritization of global problems rather than seeing inherent linkages between them.

Perhaps the best way to consider the treasure impulse in the context of technological innovation is the concept of exaptation, which was popularized by the great paleontologist Stephen Jay Gould in "Tires to Sandals," an essay that described the innovations people in developing countries make in utilizing objects that might otherwise be deemed redundant.[51] Old discarded tires can be turned into sandals, or insulation for homes, or lifesavers on boats. Similarly, the acquisition of materials can be positively channeled if all manufacturers and consumers are made to consider innovations in their production and use that go beyond the proximate purpose. At an evolutionary scale, exaptation can be a manifestation of our treasure impulse for material renewal.

For us to best utilize our abilities in seeking out such material exaptation, those who are most concerned about the environment need to have the knowledge base to embark upon such a quest. One of the most distinguished researchers on peak oil, Kenneth Deffeyes, perhaps captures the challenge that many students and concerned citizens face in understanding environmental challenges and dealing with the conundrum of human resource consumption patterns: "Understanding environmental interactions involves some of the most difficult issues in engineering, chemistry, physics and computer science. Regrettably, many environmental students are not interested in the hard stuff."[52] While this may reflect a technical elitism, there is some clear truth to the view expressed by Deffeyes. Environmental impact analysis and an understanding of resource consumption patterns are objective and highly technical enterprises.

Consequently, the foreboding of an end to human enterprise should be dismissed at least until humanity has the wherewithal to consider all of its options and an informed global society is trying in multifarious ways to improve our plight. Basic scientific understanding of ecological constraints, as well as the exploratory infusion of ideas for new technologies, must be essential parts of the ever-expansive phenomenon of education.

As we consider this prospect of galvanizing human interest in constructive consumption, I am not optimistic about a world that is identical to the one that exists right now. Indeed, I expect the world to be very different in years to come, and perhaps tougher in some ways, such as space for living and mobility, while easier in terms of disease management or information processing. What I have tried to argue is that the relationships among population, consumption of nonrenewable resources, and the environment must be considered in a larger context of human development, including both poverty alleviation and the ultimate evolutionary goal of species survival. Humans have been endowed with an impulse to seek treasures from the earth in various forms, which has served civilizations and their histories ambivalently. As we ponder our environmental predicament, it is tempting to disparage this impulse as a mark of materialism. That would be rash, considering the full spectrum of the challenges that face humanity.

All too often we seek comfort in a linear vision of the world, based on a selective view of history and science that best suits our personal and emotional proclivities. We need to consider that in a system of deliberative democracy, as well as in a world of tremendous inequalities, we will most likely have a suboptimal outcome in environmental performance. A purely utilitarian view of consumption may lead to an artless world and also one where the very impulse to discover and innovate is stifled. Let us instead embrace our impulse for treasures, while also seeking to understand the complexity of earthly elements and the materials that are the basis of all existence. In doing so, we are more likely to elude a self-inflicted ending to the story of humans and humanity.

EPILOGUE: EMBRACING UNCERTAINTY

> The future has several attributes
> For the weak, it is impossible
> For the fainthearted, it is unknown
> But for the thoughtful and valiant, it is ideal
> —Victor Hugo, *Les Misérables*, 1862

Not long ago my two children, ages ten and seven, repeatedly asked me to accompany them to a new movie in town. Consumed with the task of writing, I could scarcely find time to fit in a cinema outing with the kids to whom I was dedicating this book. My older son, Shahmir, informed me that this movie would be different from other animated features and was relevant to my interests: "It's about the environment, Abba [Dad, in Urdu]—you'll like it."

The film was *Wall-E,* an imaginatively designed satire about a resource-depleted and polluted planet Earth, where humans are forced to evacuate to live in space. A lone robot is left cleaning up the mess after a consumption-frenzied populace wallows in its woes of obesity and suicidal self-indulgence. Regardless of whether one finds this scenario to be hyperbolic, perhaps a triumph of the film was its ability to stimulate constructive debate from various ends of the political spectrum. Clearly, the ecological message was appealing to environmentalists and liberals, who hailed the movie for gently urging a "clear-headed responsibility for the health of the planet as well as one's body and mind."[1] Some on the right berated the film as Malthusian fear-mongering, but the *American Conservative*'s Patrick Ford called it a gem because it highlighted the conservative belief that "mass consumerism is not just the product of big business, but of big business wedded with big government. In fact, the two are indistinguishable in *Wall-E*'s future. The government unilaterally provided its citizens with everything they needed, and this lack of variety led to Earth's downfall."[2]

The most important audience for the movie, kids all over the world, clearly had fun watching it and maybe learned some subliminal messages about environmental planning on the way. The aim of this book has been similar: to inform a wider audience, across the political spectrum, about the importance of humanity's relations with materials; to encourage us to make connections between disparate disciplines; and to be willing to always question ourselves.

Effective planning is certainly important no matter what the outlook on mineral depletion, and both individual action and institutional change are needed. As I was completing finishing touches on this book I attended a plenary lecture at the annual meeting of the American Association for the Advancement of Science, in February 2009, by planetary geologist Susan Kieffer. This famed scientist and futurist urged participants to "celebrate the earth," while preparing for resource depletion and sudden natural crises. She reminded the audience of the supervolcanic eruption of Lake Toba in Sumatra around seventy-five thousand years ago, which wiped out most of humanity. Only around five thousand humans are believed to have survived the calamity globally. Kieffer proposed the establishment of a Center for Disaster Control for Planet Earth (CDCPE), in which scientists and policy makers would work collectively to find solutions to ecological challenges and avert sudden declines in human development from both natural and anthropogenic sources. Perhaps such an institutional response is needed at a global level, not just for specific issues like climate change that are now receiving prime attention, but collectively on natural resource usage more generally.

If we were to come up with a simple algorithm for considering global challenges, what would it be? It would be sheer hubris for me to claim to have the solutions to these problems. However, there may be a few points that we can consider in trying to relate to the challenges of consumption and the environment. A very general five-point plan of action might look something like the following.

1. Give conservation measures priority within existing frames of reference by mandating responsible ecological design that does not disparage consumption but makes it more efficient.
2. Identify proximate challenges to human development through an integrative approach that considers potential paths to livelihood creation and a global safety net for basic health services: nonrenewable resources may be part of such a development path and act as a catalyst sector.
3. Reconfigure environmental policy toward materials management that seeks to consider different scenarios of lifestyles, based on resource constraints rather than one ideal outcome.
4. Harness nonrenewable transitional materials with a clear control on the temporal scale of the operations and the speed and scale of restoration versus extraction, and the valuing of the ecological system in which these resources are situated.
5. Invigorate scientific creativity through a greater emphasis on environmental education that is premised on retrieving human-created materials from the environment and synthesis of essential materials such as fuels through biochemical innovation.

Each of these points could take the form of many books as well and have its own series of implementable steps. The goal of this book has been to harmonize many existing ideas and provide some new context for conversation on consumption and the environment through a focus on minerals as the most fundamental resource base.

The same diversity of temperaments, beliefs, and cultural displays of pageantry that we so cherish in other worldly endeavors is also manifest in our diverse responses to these environmental challenges. The ultimate environmental impact variable will always be population. Increasing life expectancy through technological innovation and attempts at achieving "immortality" regardless of conventional population control narratives will clearly have an impact on planetary resources. The only way for us to move constructively forward in such a world without callously hoping for epidemics or crises to decimate populations is to consider how best we can cultivate our treasure impulse to seek materials that have already gone through human transformation.

Having exhausted our mineral resources from earthly origins, we will perhaps need to invigorate the synthetic side of our treasure impulse. This transition will also necessitate innovative sources of energy, however, since there is no free lunch in the art and craft of material synthesis. Already we are beginning to see synthetic or "created" stones, and perhaps this is a prelude to synthetic fuels that ultimately derive their energy from more dependable sources. One story that could be the script for a chemical parable for kids is that of Robert Wentorf, Jr., who worked at General Electric's research center in Niskayuna, New York. As an avid snacker, Wentorf always kept a jar of peanut butter and some crackers on his desk. In his research to manufacture synthetic diamonds, he decided to have some fun in the process. Knowing that diamonds were essentially carbon and most edibles were carbon-based molecules, he used his favorite brand of peanut butter in the trial run for diamond synthesis. With the technological prowess at his disposal, he converted a dollop of his favorite snacking spread into tiny green crystals of diamonds! The green color, to Wentorf's amusement, was the result of some nitrogen bonding in the process. His friend Herb Strong is known to have laughed afterward and said: "The pity is you can't turn diamonds into peanut butter!"[3] Perhaps that day will come as well.

Our path toward synthetic consumption with a new range of energy sources may lead to a very different world from the one we know. That world might be closer to the fictional representation in Ridley Scott's *Blade Runner,* or Gene Roddenberry's *Star Trek* or maybe even *Wall-E,* but it will still be our world with all its material desires, quests for efficiency, and creative paths that test our minds.

APPENDIX: NOTABLE MINERALS AND THEIR HUMAN USES

Aggregates	Concrete, building construction, roads, bridges, sewer and water systems
Aluminum	Aircraft parts, automotive parts (engine blocks and cylinder heads, heat exchangers, transmission housings, and wheels), railroad cars, seagoing vessels, packaging (foil, cans, cookware), building construction (siding, windows, skylights, weatherproofing, doors, screens, gutters, downspouts, hardware, canopies, and shingles), electrical applications (overhead power lines, wires and cables), pharmaceuticals (antacids, antiperspirants), water treatment
Antimony	Alloys, flameproofing compounds, batteries, plastics, ceramics, glass, infrared detectors and diodes, cable sheathing, small arms, paints, medicine
Arsenic	Glass production, semiconductors, wood preservation, pesticides, bronzing, pyrotechnics, laser material
Asbestos	Cement building materials (roofing, cladding, pipes), heat and acoustic insulation, fire proofing
Beryllium	Structural material for high-performance aircraft, missiles, spacecraft and communication satellites, automotive parts, computer and laser technology, X-ray windows, ceramics, nuclear industry
Bismuth	Malleable irons, thermocouple material, carrier for uranium fuel in nuclear reactors, low-melting fire detection and extinguishing systems, acrylic fibers, medicine, cosmetics
Borates	Fertilizers, disinfectant, detergent, water softener, corrosion inhibitor for antifreeze, flux in brazing, ceramics, paint, coated paper, enamels, heat-resistant glass (Pyrex), pharmaceuticals, food preservatives

Cadmium	Electroplating, nuclear reactor parts, television phosphors, batteries
Chromium	Metal plating, alloys, pigments, corrosion resistance, glass and ceramics, catalyst, oxidizing agents, anodizing aluminum, tanning leather, refractory products
Clays	Bricks, ceramics, nutritional additives, concrete, mortar
Coal	Electricity generation, steel making, chemical manufacture, production of liquid fuels, plastics, and polymers
Cobalt	Super alloy (used in jet engines and gas turbine engines), magnets, stainless steel, electroplating, batteries, cemented carbides (hard metals) and diamond tools, catalysts, pigments, radiotherapeutic agent
Copper	Building construction (wire, cable, plumbing and gas tubing, roofing and climate control systems), aircraft parts (undercarriage components, aeroengine bearings, display unit components, and helicopter motor spindles), automotive parts (wire, starter motor, bearings, gears, valve guides), industrial applications and machinery (tools, gears, bearings, turbine blades), furniture, coins, crafts, clothing, jewelry, artwork, musical instruments
Dolomite	Building stone, nutritional additives
Feldspar	Glass, ceramics, enamel, tile glazes, source of alkalis and alumina in glazes, paint, plastics, mild abrasives, welding electrodes
Fluorspar	Steel making, aluminum, fluorocarbons (used in refrigerants, blowing agents, solvents, aerosols, sterilants, fire extinguishers)
Gallium	Compound semiconductors in mobile phones, glass and mirror coatings, transistors
Germanium	Semiconductors, infrared imaging and detector systems, optical fibers, phosphor in fluorescent lamps, catalyst, radiation detectors, lasers and light detectors, medical and biological uses
Gold	Jewelry, electronics, dentistry, decorative painting of costume jewelry, watchcases, pens and pencils, spectacle frames and bathroom fittings, decoration of china and glass, store of value
Graphite	High-temperature lubricants, brushes for electrical motors, brake and friction linings, battery and fuel cells, pencil fillings, seals and gaskets, conducting linings on cables, antistatic plastics and rubbers, heat exchanger, electrodes, apparatuses and linings for the chemical industry
Gypsum	Building construction (plasterboard, plaster and cement), agriculture, glass, chemicals
Iron	Steel making, alloys
Kaolin	Filler for paper, rubber, plastic, paint and adhesives, refractories, ceramics, fiberglass, cement, catalyst for petroleum refining
Lead	Batteries, cable sheathing, lead crystal, solder and radiation protection, anti-knock compound in gasoline, plumbing, ammunition

Limestone	Aggregate, cement, fertilizer, soil conditioner, iron flux, paints, plastics, livestock feed
Lithium	Lubricants, glass and ceramics, lithium carbonate (used for aluminum reduction, batteries, pharmaceuticals), high-performance alloys for aircraft, carbon dioxide absorber in spacecraft, nuclear applications
Magnesite	Agricultural fertilizer, refractory bricks, filler in plastics and paints, nuclear reactors and rocket engine nozzles, manufacture of Epsom salts, magnesia, cosmetics, insulating material and disinfectant, fire retardant
Magnesium	Alloys used for aircraft, car engine casings, and missile construction, refractory materials, agriculture (feed and fertilizer), filler in paper, paints, and plastics, automobiles and machinery, ceramics, fire retardant, pyrotechnics and flares, reducing agent for the production of uranium and other metals from their salts
Manganese	Steel making, alloys, batteries, colorants and pigments, ferrites, welding fluxes, agriculture, water treatment, hydrometallurgy, fuel additives, oxidizing agents, odor control, catalysts, sealants, metal coating, circuit boards
Mercury	Thermometers, barometers, diffusion pumps, electrical apparatus, electrodes, batteries, chlorine and sodium hydroxide manufacture, lighting, pesticides, dentistry
Molybdenum	Alloys, catalyst in petroleum refining, heating elements, lubricants, nuclear energy applications, missile and aircraft parts, electrical applications
Nickel	Stainless steel, corrosion-resistant alloys, gas turbines, rocket engines, plating, coins, catalysts, burglar-proof vaults, batteries
Niobium	Alloys, stainless steels, advanced engineering systems (space programs), nuclear industry, electrical products, jewelry
Palladium	Jewelry, watches, surgical instruments, catalysts, dentistry (crowns), electrical contacts, hydrogen gas purification
Phosphate rock	Fertilizers, detergents, flame retardants, food and beverages, animal feeds, metal treatment, water treatment, pulp and paper, glass and ceramics, textiles and synthetic fibers, plastics, rubber, pharmaceuticals, cosmetics, petroleum production and products, construction, pesticides, toothpaste, mining, leather, paints, fuel cells
Phosphorus	Safety matches, pyrotechnics, incendiary shells, smoke bombs, tracer bullets, glass, calcium phosphate (used to produce fine chinaware), steel making, cleaning agent, water softener, pesticides
Platinum	Jewelry, coins, auto catalysts, electronics, glass, dentistry, chemical and electrochemical catalysts, petroleum, laboratory equipment, antipollution devices in cars, investment, anti-cancer drugs, implants (pacemakers, replacement valves)

Plutonium	Nuclear fuel and weapons, pacemakers
Potash	Fertilizer, soap and detergents, glass and ceramics, chemical dyes and drugs, food and beverages
Pumice	Construction, stonewashing in textile industries, glass and metal polishing, dental supplies and paste, agriculture, sport and leisure facilities, cosmetics
Rhodium	Alloys (used for furnace windings, thermocouple elements, bushings for glass fiber production, electrodes for aircraft spark plugs, laboratory crucibles), electrical contact material, optical instruments, jewelry, industrial catalysts, automotive catalytic converters
Sand and gravel	Concrete, bricks, roads, building materials
Selenium	Photoreceptors (used in the manufacture of plain paper photocopiers and laser printers), electronic applications, glass, pigments, alloys, biological applications, rubber, lubricants, catalysts
Silica	Glass (bottles and jars)
Silver	Photography (X-ray film for medical, dental, industrial uses), jewelry, electrical applications, batteries, solder and brazing alloys, tableware, mirrors and glass, coins
Soda ash	Glass, detergents, chemicals, water treatment, flue gas desulfurization, pulp and paper
Sulfur	Sulfuric acid, ammunition, fungicide, vulcanization of natural rubber
Talc	Paper, plastics, paints, ceramics, refractories, roofing, rubber, cosmetics, pharmaceuticals, agrochemical, animal feed, cement, glass fiber
Tantalum	Electrolytic capacitors, alloys (used in aircraft and missile manufacture), lining for chemical and nuclear reactors, wires, surgery (used in sutures and as cranial repair plates), cameras
Tin	Tinplates, alloys, solder, pewter, chemicals, panel lighting, frost-free windshields
Titanium	Production of lightweight alloys, aircraft components (jet engines, aircraft frames), automotive components, joint replacement (hip ball and sockets), paints, watches, chemical processing equipment, marine equipment (rigging and other parts exposed to seawater), pulp and paper processing equipment, pipes, jewelry
Tungsten	Alloys (used in filaments for electric lamps, electron and television tubes, metal evaporation work), ammunition, chemical and tanning industry, paints, X-ray targets
Uranium	Nuclear fuel, nuclear weapons, X-ray targets, photographic toner
Vanadium	Alloys (especially in steel), catalysts, pigments for ceramics and glass, batteries, medical compounds, pharmaceuticals, electronics

Zinc	Galvanizing, alloys, brass, batteries, roofing, water purification, coins, zinc oxide (used in manufacture of paints), rubber products, cosmetics, pharmaceuticals, floor coverings, plastics, printing inks, soap, textiles, electrical equipment, ointments, zinc sulfide (used in making luminous dials, X-ray and TV screens, paints, fluorescent lights)
Zirconium	Ceramics, refractories, foundry sands, glass, chemical piping in corrosive environments, nuclear power reactors, hardening agent in alloys, heat exchangers, photographic flashbulbs, surgical instruments

Source: International Institute for Environment and Development, *Breaking New Ground: Mining, Minerals, and Sustainable Development* (London: Earthscan, 2003)

NOTES

Introduction: Alchemy of a Material World

1. For an excellent history of Leadville and the community's remarkable efforts to preserve its heritage and revitalize it economically, see Gillian Kluckas, *Leadville: The Struggle to Revive an American Town* (Washington, D.C.: Island, 2004).

2. Details of nineteenth-century life in Leadville are derived from Beth Sanstetter and Bill Sanstetter, *The Mining Camp Speaks* (Denver, Colo.: Benchmark, 1998).

3. For a history of elemental experimentation, see Richard Morris, *The Last Sorcerers: From Alchemy to the Periodic Table* (New York: Joseph Henry, 2005). Another readable account of this history is Trevor Levere, *Transforming Matter: A History of Chemistry from Alchemy to the Buckyball* (Baltimore: Johns Hopkins University Press, 2001).

4. The most authoritative history of the periodic table is probably Eric Scerri, *The Periodic Table: Its History and Significance* (New York: Oxford University Press, 2006).

5. An academic analysis of this issue is provided in R. B. Gordon et al., "Metal Stocks and Sustainability," *Proceedings of the National Academy of Sciences* 103, no. 5 (2006).

6. Eve Curie, *Madame Curie: A Biography* (New York: Da Capo, 2001), 156.

7. L. D. Anderson, "Geopolitics and the Cold War Environment: The Case of Chromium," *GeoJournal* 37, no. 2 (October 1, 1995): 209–214.

8. Estimates from the American Chemical Society in various popular publications. For an excellent journalistic account of human ingenuity in harnessing these elements and compounds into products for our daily lives, see Ivan Amato, *Stuff: The Materials the World Is Made Of* (New York: Basic, 1997).

9. Charles Moore, "Trashed: Across the Pacific Ocean, Plastics, Plastics Everywhere," *Natural History,* November 2003.

10. Gecia Matos and Lorie Wagner, *Consumption of Materials in the United States, 1900–1995* (Washington, D.C.: U.S. Geological Survey, 1997).

11. Gary Gardner, *Mind over Matter: Recasting the Role of Materials in Our Lives* (Washington, D.C.: Worldwatch Paper no. 144, 1998).

Chapter 1. What Lies Beneath

1. For an informative history of human burial rituals and the underlying rationales for these practices, see Penny Coleman, *Corpses, Coffins, and Crypts: A History of Burial* (New York: Henry Holt, 1997).

2. For highly readable accounts of this concept, see Steven Johnson, *Emergence: The Connected Lives of Ants, Brains, Cities, and Software* (New York: Scribner, 2002), and Stuart A. Kauffman, *The Origins of Order: Self-Organization and Selection in Evolution* (New York: Oxford University Press, 1993).

3. James P. Ferris, "Montmorillonite-Catalyzed Formation of RNA Oligomers: The Possible Role of Catalysis in the Origins of Life," *Philosophical Transactions of the Royal Society B* 361 (2006): 1777–1786.

4. Madeline Vargas et al., "Microbiological Evidence for Fe(III) Reduction on Early Earth," *Nature* 395 (1998): 65–67.

5. C. Boesch and H. Boesch, "Mental Map in Wild Chimpanzees: An Analysis of Hammer Transports for Nut Cracking," *Primates* 25 (1984): 160–170. For a more integrative study see William McGrew, *Chimpanzee Material Culture* (Cambridge: Cambridge University Press, 1992).

6. Gad Saad, *The Evolutionary Bases of Consumption* (Mahwah, N.J.: Lawrence Erlbaum, 2007).

7. The causal relation between environmental degradation and the collapse of societies, particularly Easter Island, has been contested in recent scholarship. However, there is little doubt that environmental decline, whether through overuse or introduction of alien species, contributed to the process. For a review of the debate on Easter Island, see Jared Diamond, "Easter Island Revisited," *Science* 317 (2007): 1692–1694.

8. The seminar was organized by archaeologist Patricia McAnany and anthropologist Norman Yoffee. For an excellent review of this seminar, see George Johnson, "A Question of Blame when Societies Fall," *New York Times*, December 25, 2007. For a more nuanced view of civilizational collapse, see Joseph Tainter, *The Collapse of Complex Societies* (Cambridge: Cambridge University Press, 1990).

9. For a detailed study of the Swaziland cave since its discovery, see R. A. Dart and P. Beaumont, "Amazing Antiquity of Mining in Southern Africa," *Nature* 216 (1967): 407–408. Also refer to the following for research on some of the earliest accounts of quarrying: P. M. Vermeersch and E. Paulissen, "The Oldest Quarries Known: Stone Age Miners in Egypt," *Episodes* 12, no. 1 (March 1989): 35–36; Alan Walker and Pat Shipman, *The Wisdom of the Bones* (New York: Alfred Knopf, 1996).

10. Cyril Stanley Smith, "Into the Smelting Pot," a review of R. F. Tylecote's *A History of Metallurgy, Times Literary Supplement*, November 4, 1977, 1301.

11. Mark Aldenderfer, Nathan M. Craig, Robert J. Speakman, and Rachel Popelka-Filcoff, "Four-Thousand-Year-Old Gold Artifacts from the Lake Titicaca Basin, Southern Peru," *Proceedings of the National Academy of Sciences*, March 31, 2008.

12. Richard Rudgeley, *The Lost Civilizations of the Stone Age* (New York: Free Press, 2000).

13. There are numerous books on the history of these various metallic ages. Among the classic works are Theodore A. Wertime, *The Coming of the Age of Steel* (Chicago: University of Chicago Press, 1962); Theodore A. Wertime and James D. Muhly, *The Coming of the Age of Iron* (New Haven: Yale University Press, 1980); Kristian Kristiansen and Thomas B. Larsson, *The Rise of Bronze Age Society: Travels, Transmissions, and Transformations* (Cambridge: Cambridge University Press, 2006).

14. F. L. Koucky and A. Steinberg, "Ancient Mining and Mineral Dressing on Cyprus," in *Early Pyrotechnology*, edited by T. Wertime and S. Wertime (Washington, D.C.: Smithsonian Institution Press, 1982).

15. Quotations on smelting taken from Eugenia W. Herbert, *Iron, Gender, and Power* (Bloomington: Indiana University Press, 1993). These quotations were originally written in French and translated by Herbert.

16. Bade Ajuwon, "Ogun's Iremoje: A Philosophy of Living and Dying," in *Africa's Ogun*, ed. Sandra Barnes (Bloomington: Indiana University Press, 1997).

17. Eugenia Herbert, *Red Gold of Africa* (Madison: University of Wisconsin Press, 1984).

18. For a detailed account of this period, see D. H. Fischer, *The Great Wave: Price Revolutions and the Rhythm of History* (New York: Oxford University Press, 1999). Also refer to Daniel Roche, *A History of Everyday Things: The Birth of Consumption in France*, translated by Brian Pearce (Cambridge: Cambridge University Press, 2000).

19. For an interesting exposition of real-world Indiana Jones stories and a general discussion of the importance of religion in society, see Steven Sora, *Treasures from Heaven: Relics from Noah's Ark to the Shroud of Turin* (New York: John Wiley and Sons, 2005).

20. Quoted in Kathryn Taylor Morse, *The Nature of Gold: An Environmental History of the Klondike Gold Rush* (Seattle: University of Washington Press, 2003), 194–195. This theme is also found in the writings of Frederick Jackson Turner on the significance of the frontier in American history; Frederick Jackson Turner, *Rereading Frederick Jackson Turner: "The Significance of the Frontier in American History" and Other Essays* (New Haven: Yale University Press, 1999).

21. Peter Bernstein, *The Power of Gold: The History of an Obsession* (New York: John Wiley and Sons, 2000), 73.

22. For a phenomenal history of salt see Mark Kurlansky, *Salt: A World History* (New York: Penguin, 2004).

23. Thomas Jefferson also considered the practice in America in the late eighteenth century. Details about desalination history are considered in Giorgio Nebbia, *A Short History of Water Desalination* (Rome: Azienda Grafica Italiana, 1966). Also refer to the following Web site, titled Encyclopaedia of Desalination and Water Resources: www.desware.net/.

24. Madeline Zelin, *The Merchants of Zigong: Industrial Entrepreneurship in Early Modern China* (New York: Columbia University Press, 2008).

25. Marq de Villiers and Sheila Hirtle, *Timbuktu: The Sahara's Fabled City of Gold* (New York: Walker, 2007). For East African salt trade see Paul E. Lovejoy, *Salt of the Desert Sun: A History of Salt Production and Trade in the Central Sudan* (Cambridge: Cambridge University Press, 2003).

26. Kurlansky, *Salt*, 164.

27. Quoted in Stanley Wolpert, *India* (Berkeley: University of California Press), 204.

28. Quoted in Tahir Shah, *In Search of King Solomon's Mines* (London: Arcade, 2003), 9. Shah's book is an entertaining and informative travelogue through the Horn of Africa with some startling revelations about gold mining in this day and age as well.

29. Ranjit S. Dighe, *The Historian's Wizard of Oz* (Westport, Conn.: Greenwood, 2002).

30. Crystals arguably have properties that make them comparable to life forms—unique structures that can absorb materials and energy (often forming beautifully latticed forms like a snowflake) and the ability to propagate through molecular adhesion. They are of course not alive as an organism is, but several researchers are studying crystals to understand how molecular organization may have led to the emergence of the first living cells. For an excellent multimedia exposition of various theories on the origin of life, see the History Channel's special program *How Life Began* (produced in 2008), featuring the work in Robert Hazen, *Genesis: The Scientific Quest for Life's Origins* (Washington, D.C.: Joseph Henry Press of the U.S. National Academies of Sciences, 2005).

31. Flora Peschek-Bohmer and Gisela Screiber, *Healing Crystals and Gemstones* (Old Saybrook, Conn.: Konecky and Konecky, 2002).

32. For a fabulous account of the history of jade in Chinese folklore, with a particular focus on the obsession of nineteenth-century Chinese, Burmese, and Europeans with the stone, see Adrian Levy and Cathy Scott-Clark, *The Stone of Heaven: Unearthing the Secret History of Imperial Green Jade* (Boston: Back Bay, 2003).

33. The fable of "stone soup," in which cooperation among villagers is induced by enticing each to contribute one ingredient to a pot of shared soup with only a bland stone to start has been told in many societies. The town of Almeirim in Portugal lays claim to the story and has restaurants galore that claim to serve the ultimate recipe of *sopa de pedra*.

34. For an academic account of the culture of consumption in the past three hundred years, see John Benson and Laura Ugolini, *Cultures of Selling: Perspectives on Consumption and Society Since 1700* (Burlington, Vt.: Ashgate, 2006). Also refer to Peter Stearns, *Consumerism in World History: The Global Transformation of Desire* (London: Routledge, 2006).

Chapter 2. Creating Value

1. Details of this remarkable discovery are chronicled in J. Garcia-Guinea and J. M. Calaforra, "Mineral Collectors and the Geological Heritage: Protection of a Huge Geode in Spain," *European Geologist Magazine* 1 (2001): 4–7.

2. For a stunning account of geodes worldwide, see Brad Lee Cross and June Culp Zeitner, *Geodes: Nature's Treasures* (San Diego: Gem Guides, 2006).

3. The power of minerals pervades the comic-strip world, most prominently in the form of Superman's origins and nemesis. The planetary city of his birth on Krypton is depicted as a crystalline fortress governed by an intricate balance of energy emanating from gemlike crystal forms. The superhero arrives on Earth surrounded by life-giving shards of crystals in his space vessel. His ultimate fortress of protection on the North Pole is also constructed of crystalline structures of various kinds that slowly reveal his destiny. The negative power of the elements is also manifest in mineral form through that most dreaded of all green creations—kryptonite. Fictional as this may be, there is indeed an element called krypton—an inert gas with a name that may have inspired the writers—though the chemical formula is referenced in the movies as "sodium lithium boron silicate hydroxide." In 2007 a mineral with such a formula was discovered in a Serbian mine near the town of Jadar, though it was quite inert and white rather than green. For a highly readable history of the man from Krypton, see Les Daniels, *Superman: The Complete History* (New York: Chronicle, 2004).

4. Preliminary analyses of the formation of the crystals and the technical data gathered from the site are provided in J. Garcia-Guinea et al., "Formation of Gigantic Gypsum Crystals," *Journal of the Geological Society, London* 159 (2002): 347–350.

5. The story of Indra's net is now more commonly associated with the Mahayana Buddhist school and is usually dated to the third-century *Avatamsaka Sutra*. The Chinese Huayan school popularized the metaphor further, though the Hindu roots are undeniable since Indra is a Hindu deity dating to the *Rig Veda* (1700–1100 B.C.). For a detailed account of this fascinating metaphor see Francis Cook, *Hua-Yen Buddhism: The Jeweled Net of Indra* (College Park: Pennsylvania State University Press, 1977).

6. Gem and Jewelry Export Promotional Council of India, statistics quoted in a promotional video produced by the group in 2004.

7. Nick Robins, *The Corporation that Changed the World: How the East India Company Shaped the Modern Multinational* (London: Pluto, 2006). Accounts of gem-seeking travels can be found in Kevin Rushby, *Chasing the Mountain of Light* (London: Robinson, 2004); Omar Khalidi, *Romance of the Golconda Diamonds* (New York: Mapin, 2006); Bernhard Graf, *Gems: The World's Greatest Treasures and Their Stories* (Prestel, 2001).

8. The diamond discovery in Brazil is credited to Friar Bernardo da Fonseca, a resident of former Portuguese India who found some specimens north of Rio de Janeiro in 1727. This region was given the name Diamantina. Thousands of small-scale miners, or *gareimperos*, scoured the landscape, and by the end of the eighteenth century, Brazil was producing four times as many diamonds as India. The early accounts of these miners (*scravos do diamante*—slaves of the diamond) in Brazil are best narrated in Richard Burton, *Explorations of the Highlands of Brazil, with a Full Account of the Gold and Diamond Mines*, vol. 1 (New York: Greenwood, 1969 [reprint]), and John Mawe, *Travels Through the Diamond District of Brazil* (London: Longman Hurst, 1812).

9. George E. Harlow, ed., *The Nature of Diamonds* (Cambridge: Cambridge University Press, 1998).

10. Quoted in Kevin Krajick, *Barren Lands: An Epic Search for Diamonds in the North American Arctic* (New York: Times, 2001).

11. World Diamond Council data: www.diamondfacts.org (accessed February 1, 2008).

12. See for example Matthew Hart, *Diamond: The History of a Cold-Blooded Love Affair* (New York: Plume, 2001).

13. For an excellent geographical analysis of diamonds and conflict linkages in Africa, see Philippe LeBillon, "Diamond Wars? Conflict Diamonds and Geographies of Resource Wars," *Annals of the Association of American Geographers* 98 (2008): 345–372. Also see Richard Snyder and Ravi Bhavnani, "Diamonds, Blood, and Taxes: A Revenue-Centered Framework for Explaining Political Order," *Journal of Conflict Resolution* 49, no. 4 (August 1, 2005): 563–597.

14. There is a profusion of books on the topic of diamonds, conflict, and social injustice. They all need to be read carefully to ascertain facts from anecdotal speculation. While the overall concerns about diamonds and conflict are valid, the hysteria created by some of the accounts often misses the larger issues of development that are associated with diamonds. For the more vitriolic accounts against the industry, see Janine Roberts, *Glitter and Greed: The Secret World of the Diamond Empire* (New York: Disinformation, 2007); Tom Zoellner, *The Heartless Stone: A Journey Through the World of Diamonds, Deceit, and Desire* (New York: Macmillan, 2006). Also see Leo P. Kendall, *Diamonds Famous and Fatal: The History, Mystery, and Lore of the World's Most Famous Gem* (New York: Barricade, 2001).

15. Several reports by expert panels were commissioned on the linkages between resource extraction and conflict in Africa by the U.N. Security Council. These are available for download from a Web site prepared by Columbia University's Center for International Organization: www.securitycouncilreport.org.

16. Douglas Farah, *Blood from Stones: The Secret Financial Network of Terror* (New York: Broadway, 2004). Since the publication of this book Farah has shifted his interest to studying the arms trade and the role played by the notorious Russian arms trader Viktor Bout, in Douglas Farah and Stephen Braun, *Merchant of Death: Money, Guns, Planes, and the Man Who Makes War Possible* (New York: Wiley, 2008).

17. Ibrahim Warde, *The Price of Fear: The Truth Behind the Financial War on Terror* (Berkeley: University of California Press, 2007).

18. National Commission on Terrorist Attacks, *The 9/11 Commission Report: Final Report of the National Commission on Terrorist Attacks upon the United States* (New York: W. W. Norton, 2004), 171.

19. Data on Botswana from CIA World Factbook, 2008, online at https://www.cia.gov/library/publications/the-world-factbook/geos/bc.html#Intro.

20. Data on unemployment in Botswana are somewhat unreliable. This figure is based on the CIA World Factbook estimates from 2004.

21. John D. Holm, *Diamonds and Distorted Development in Botswana* (Washington, D.C.: Center for Strategic and International Studies, Africa Policy Forum, January 8, 2007).

22. *The Ecologist* 34 (2004): 14–15.

23. J. Clark Leith, *Why Botswana Prospered* (Montreal: McGill-Queen's University Press, 2006).

24. Census of India data accessed online (last census was in 2001), February 12, 2007, www.censusindia.gov.in/.

25. The phenomenon of "market-dominant minorities," such as Lebanese gem traders, Jewish and Jain cutters and polishers, or Parsi and Zoroastrian merchants, has been studied in considerable detail in Amy Chua, *World on Fire* (New York: Anchor, 2003). Her analysis suggests that globalization has tended to favor the networks of these market-dominant minorities, leading to resentment in local populations and thereby sparking conflict. Chua's initial argument regarding the dominance of certain ethnic minorities, such as the Chinese in the Philippines, is compelling. Her chapter on Jews in Russia is, however, less convincing because she does not try to answer the basic question of how this dominance occurred. In Russia, there was extreme discrimination against the Jews and still they were able to rise to the top (unlike some of her other cases of market-dominant minorities). Some of Chua's proposals for reform toward the end of the book, such as equity markets in which the poor have a direct share, are worth considering. However, this can all be handled more dialectically with democratic reforms taking place—so long as the rule of law prevails. For example, the Rwandan genocide's proximate cause was primarily institutional collapse of law enforcement, either domestically or via international channels. Transition to markets and democracy can occur without violence, as Chua rightly points out, as in the relatively positive case of Thailand.

26. Quoted in Hart, *Diamond*, 233.

27. Geshe Michael Roach, *The Diamond Cutter: The Buddha on Managing Your Business and Your Life* (New York: Doubleday, 2003), 29. Although Michael Roach has been ordained as a geshe (high priest or an equivalent), he has also been reprimanded by the Dalai Lama because he does not abide by many of the traditional Buddhist vows for a monk, such as chastity and short hair. Roach managed to convert a $50,000 investment into a huge jewelry retailer called Andin International. He later sold this investment and started various philanthropies, including Diamond Mountain University, a meditation school in Arizona where he is now based.

28. The word "carat" has a varied etymology in Greek (*keration*), Arabic (*qirat*), and Italian (*carato*). The weight itself is tied to the usual weight of the fruit from the leguminous carob tree, which is also called a carat—every single pod of which weighs approximately one-fifth of a gram.

29. For a detailed review of child labor in Surat, see Kiran Desai and Nikhil Raj, *Child Labour in Diamond Industry of Surat* (Noida, India: V. V. Giri National Labour Institute, 2001).

30. Nicholas D. Kristof, "Wretched of the Earth," *New York Review of Books* 54, no. 9 (May 31, 2007).

31. Some researchers do not consider amber fossilized because the word implies the replacement of organic material with minerals over time, whereas amber is all organic material (mainly terpene compounds) that has been polymerized and hardened.

32. David A. Grimaldi, *Amber: Window to the Past* (Washington, D.C.: Smithsonian, 2003).

33. Michael Wines, "Yantarny Journal: Russians Awaken to a Forgotten SS Atrocity," *New York Times,* January 31, 2000.

34. R. J. Cano, Gatesy DeSalle, W. Wheeler, and D. Grimaldi, "DNA Sequences from a Fossil Termite in Oligo-Miocene Amber and Their Phylogenetic Implications," *Science* 257 (1992): 1933–1936.

35. R. J. Cano and M. K. Borucki, "Revival and Identification of Bacterial Spores in 25- to 40-Million-Year-Old Dominican Amber," *Science* 268 (1995): 1060–1064. There was also a controversial assertion that bacteria preserved in salt crystals had been isolated by another team five years later; Russell H. Vreeland, William D. Rosenzweig, and Dennis W. Powers, "Isolation of a 250-Million-Year-Old Halotolerant Bacterium from a Primary Salt Crystal," *Nature* 407, no. 6806 (October 19, 2000): 897–900.

36. Neil H. Landman and Paula Mikkelsen, *Pearls: A Natural History* (New York: Harry N. Abrams, 2001).

37. Surah 22, verse 23 of the *Quran* reads: "Allah will admit those who believe and work righteous deeds, to Gardens beneath which rivers flow: they shall be adorned therein with bracelets of gold and pearls; and their garments there will be of silk."

38. An example of such a pearl, congealed to the shell of the oyster, can be seen in the collections of the Field Museum of Chicago.

39. Hughes Edwards, *Port of Pearls* (Kalamunda, Western Australia: Tangee, 1984).

40. M. J. Hibbard, *Mineralogy: A Geologist's Point of View* (New York: McGraw-Hill, 2002), 116.

41. James Shigley, Gemological Institute of America, Carlsbad, California, personal communication, February 25, 2008.

42. For an absorbing account of the science and industry of synthetic diamonds, see Robert M. Hazen, *The Diamond Makers,* rev. ed. (Cambridge: Cambridge University Press, 1999).

43. Joshua Davis, "The Diamond Wars Have Begun," *Wired,* September 2003.

44. Data collected by Mary Ackley, research assistant to the author. Diavik reported fuel consumption in liters per carat, so an average density for diesel fuel of 850 grams per liter was used for conversions.

45. De Beers data reported by the company in its public reports in different units and converted by the author: 0.289 gigajoules per carat, converted to kilowatt-hours per carat. All other data collected by the author directly from contacts with sources at companies.

46. Details about the company at Life-Gem's Web site, www.lifegem.com. An interview with the founder of the company, Greg Herro, on National Public Radio's *Weekend*

Edition, November 9, 2002, can be found online at www.npr.org/templates/story/story.php?storyId=837580.

Chapter 3. The Rush Factor

1. Paul Bairoch, *Cities and Economic Development: From the Dawn of History to the Present* (Chicago: University of Chicago Press, 1991).

2. Jane Jacobs, *The Economy of Cities* (New York: Vintage, 1970).

3. Jacobs, *Economy of Cities,* 23.

4. Vere Gordon Childe, *The Dawn of European Civilization* (New York: Random House, 1958).

5. For a detailed study of precolonial sub-Saharan African cities, see Catherine Coquery-Vidrovitch, *The History of African Cities South of the Sahara: From Origins to Colonization* (Princeton: Markus Weiner, 2005).

6. Quoted in Mary Jo Ignoffo, *Gold Rush Politics: California's First Legislature* (Sacramento: California State Capitol Publications, 1999), 47. The rush of Silicon Valley entrepreneurs has also been analogized to the gold rush of Argonauts from Greek mythology who searched tirelessly for the Golden Fleece, in AnnaLee Saxenian, *The New Argonauts: Regional Advantage in a Global Economy* (Cambridge: Harvard University Press, 2007).

7. James Stanley, *Digger: The Tragic Fate of California's Indians from the Missions to the Gold Rush* (New York: Crown, 1997), 62.

8. Quoted in Robert V. Hine and John Mack Faragher, *The American West: A New Interpretive History* (New Haven: Yale University Press, 2000), 234.

9. Hine and Faragher, *The American West,* 241.

10. Leonard L. Richards, *The California Gold Rush and the Coming of the Civil War* (New York: Knopf, 2007).

11. Quoted in H. W. Brands, *The Age of Gold: The California Gold Rush and the New American Dream* (New York: Anchor, 2002), 23.

12. Quoted in Rebecca Solnit, "Winged Mercury and the Golden Calf: Two Elements, One Economic Theory, and a Cascading Torrent of Collateral Damage," *Orion* (September–October 2006), 17.

13. Paul David and Gavin Wright, "Increasing Returns and the Genesis of American Resource Abundance," *Industrial and Corporate Change* 6, no. 2 (1998): 223.

14. For an excellent discussion of Scandinavian industrialization, see Magnus Blomstrom and Ari Kokko, "From Natural Resources to High-Tech Production: The Evolution of Industrial Competitiveness in Sweden and Finland," in *Natural Resources, Neither Curse Nor Destiny,* ed. Daniel Lederman and William F. Maloney (Washington, D.C.: World Bank Publications, 2006).

15. Tony Crago, *Gold: Australia* (Newton, New South Wales, Australia: Woollahra, 2004).

16. Gavin Bridge, "Global Production Networks and the Extractive Sector: Governing Resource-Based Development," *Journal of Economic Geography* 8 (2008).

17. Dror Goldberg, "Famous Myths of Fiat Money," *Journal of Money, Credit, and Banking* (2005): 957–967.

18. J. B. De MacEdo, ed., *Currency Convertibility: The Gold Standard and Beyond* (New York: Routledge, 1996).

19. C. Murphy, "Is BP Beyond Petroleum? Hardly," *Fortune,* September 30, 2002.

20. Communities and Small Scale Mining Secretariat, estimates based on International Labor Organization studies of the informal mining sectors in Africa, Asia and South America; www.casmsite.org.

21. Gavin Hilson, ed., *The Socioeconomic Impacts of Artisanal and Small-Scale Mining in Developing Countries* (Rotterdam: A. Balkema, 2005).

22. Payal Sampat, "Scrapping Mining Dependence," in *State of the World 2003* (New York: Norton, 2003).

23. Peter Bernstein, *The Power of Gold: The History of an Obsession* (New York: John Wiley & Sons, 2000).

24. Bernstein, *The Power of Gold,* 230.

25. United Nations Industrial Development Organization (UNIDO), Artisanal mining and mercury usage project, funded by the Global Environmental Facility (under the Japanese Trust Fund of the World Bank), 2001–2004.

26. T. Green, *The World of Gold* (New York: Rosendale, 1993).

27. Geoffrey Wheatcroft, *The Randlords: The Exploits and Exploitations of South Africa's Mining Magnates* (New York: Atheneum, 1986).

28. R. Ally, *Gold and Empire: The Bank of England and South Africa's Gold Producers* (Johannesburg: Witwatersrand University Press, 1994).

29. H. Bhattacharya, "Deregulation of Gold in India" (London: World Gold Council Research Study No. 27, 2002).

30. V. Oldenburg, *Dowry Murder: The Imperial Origins of a Cultural Crime* (New York: Oxford University Press, 2002).

31. Quoted in Richard Manning, *One Round River: The Curse of Gold and the Fight for the Big Blackfoot* (New York: Henry Holt, 1998), 140.

32. For a discursive though somewhat dense history of fire in human societies, see Stephen J. Pyne, *Fire: A Brief History* (Seattle: University of Washington Press, 2001).

33. Si Posen, an expert at the China Coal Information Institute, quoted in Wang Ying and Winnie Shu, "Around the Markets: Future for Coal Is Brighter," *International Herald Tribune,* April 16, 2007.

34. Kenneth Pomeranz, *The Great Divergence: Europe, China, and the Making of the Modern World Economy* (Princeton: Princeton University Press, 2000).

35. The rise of whaling in the nineteenth century and its fall have been studied in detail from the perspective of economic history in terms of productivity changes spurred by technological innovation in Lance E. Davis, Robert E. Gallman, and Karin Gleiter, *In Pursuit of Leviathan: Technology, Institutions, Productivity, and Profits in American Whaling, 1816–1906* (Chicago: University of Chicago Press, 1997). It is important to note that apart from whale oils, one of the most prized products of whaling was an intestinal secretion that was used as a binding agent in perfumery. Whale meat was also

prized (the tongue being a particular delicacy across Europe, where the Basque had mastered the art of whaling centuries earlier).

36. Sven Wunder, *Oil Wealth and the Fate of the Forest* (London: Routledge, 2003).

37. James Howard Kunstler, *The Long Emergency: Surviving the End of Oil, Climate Change, and Other Converging Catastrophes of the Twenty-First Century* (New York: Grove, 2006); Peter W. Huber and Mark P. Mills, *The Bottomless Well: The Twilight of Fuel, the Virtue of Waste, and Why We Will Never Run Out of Energy* (New York: Basic, 2006).

38. In this regard the authors target Jeremy Rifkin's book *The End of Work*, calling him "a techno-dystopian" whose dire predictions of mechanization leading to mass unemployment have not materialized because of human creativity to invent new professions.

39. Several notable articles resulted from these visits, including Sheila McNulty, "Green Leaves, Black Gold," *Financial Times Magazine,* December 15–16, 2007, and Elizabeth Kolbert, "Unconventional Crude," *New Yorker,* November 12, 2007.

40. Katherine Bourzac, "Dirty Oil," *Technology Review,* January 2006.

41. The correspondence of Karl A. Clark from 1920 to 1949 regarding the oil-sands developments has been preserved by the University of Alberta and published with annotations as a book; Mary Clark Sheppard, *Oil Sands Scientist: The Letters of Karl A. Clark, 1920–1949* (Edmonton: University of Alberta Press, 1989).

Chapter 4. The Darker Side of Fortune

1. Dopamine's medical properties were discovered by the Swedish chemist Arvid Carlsson in 1951, for which he received the Nobel Prize in Medicine several decades later, in 2000. For a review of the importance of dopamine in neuroscience, see Alison Abbott, "Neuroscience: The Molecular Wake-Up Call," *Nature* 447 (2007): 368–370. For an excellent account of the chemistry of craving, see Ronald A. Ruden, *The Craving Brain* (New York: Harper Paperbacks, 2000).

2. GABA is a remarkably simple molecule with four carbon atoms, a couple of oxygens, a nitrogen, and a balance of nine accompanying hydrogen atoms.

3. Anne Becker, "Green Tea on the Brain," *Psychology Today,* June 10, 2003. For a review of the technical literature on tea's beneficial impact, see J. Bryan, "Psychological Effects of Dietary Components of Tea: Caffeine and L-Theanine," *Nutrition Reviews* 66, no. 2 (February 2008): 82–90.

4. While there is a pathological side to addiction, the propensity for excessive behavior exists in all individuals to some degree, depending on circumstantial triggers. The *Diagnostic and Statistical Manual of Mental Disorders* (4th edition, or DSM-IV) has a category called "impulse control disorders."

5. The term "meme" is derived from the Greek word *memos,* referring to mimicry. Dawkins coined it because of its phonetic similarity to gene.

6. For an excellent review of contending literature on "memetics," see Robert Aunger, *Darwinizing Culture: The Status of Memetics as a Science* (Oxford: Oxford University Press, 2001).

7. Richard Brodie, *Virus of the Mind: The New Science of the Meme* (San Francisco: Integral, 2004), 14.

8. John de Graaf, David Wann, and Thomas H. Naylor, *Affluenza: The All-Consuming Epidemic* (New York: Berrett-Koehler, 2005).

9. Jeremy Prestholdt, *Domesticating the World: African Consumerism and the Genealogies of Globalization* (Berkeley: University of California Press, 2008), 43. Concern over gluttony as well as avarice as "sins" is found in numerous cultural traditions, including in Christianity (as they are among the seven deadly sins).

10. Mihaly Csikszentmihalyi and Eugene Halton, *The Meaning of Things: Domestic Symbols and the Self* (New York: Cambridge University Press, 1981); Mihaly Csikszentmihalyi, "The Costs and Benefits of Consuming," *Journal of Consumer Research* 27 (2000): 267–272.

11. James Gleick, *Chaos: Making a New Science* (New York: Penguin, 1988), 304.

12. William M. Schaffer, "Chaos in Ecological Systems: The Coals that Newcastle Forgot," *Trends in Ecological Systems* 1 (1986): 63.

13. William M. Schaffer and Mark Kot, "Do Strange Attractors Govern Ecological Systems?" *BioScience* 35 (1985): 349.

14. J. S. Metcalfe, "Consumption, Preferences, and the Evolutionary Agenda," *Journal of Evolutionary Economics* 11 (2001): 37–58.

15. Staffan B. Linder, *The Harried Leisure Class* (New York: Columbia University Press, 1970).

16. For a detailed analysis of the issue of consumer choice and its impact on human behavior, see Barry Schwartz, *The Paradox of Choice: Why More Is Less* (New York: Harper Perennial, 2005). The anthropologist Marshall Sahlins studied consumerism in human societies and how the "culture of capitalism" was alien to tribal populations, such as the Huron Indians. They considered a desire for material things a disease, and those afflicted were isolated from the community and treated by satiating them with goods until they got the desire out of their systems. Marshall Sahlins, "The Sadness of Sweetness: The Native Anthropology of Western Cosmology," *Current Anthropology* 37, no. 3 (June 1996): 395–428.

17. Mihaly Csikszentmihalyi and Barbara Schneider, *Becoming Adult: How Teenagers Prepare for the World Of Work* (New York: Basic, 2001).

18. Thomas Friedman, *The Lexus and the Olive Tree: Understanding Globalization* (New York: Farrar, Straus and Giroux, 2000). For an excellent review of the social research surrounding the McDonald's phenomenon that is often termed "glocalization," and is considered a safe response to a risk-averse society, see Bryan S. Turner, "McDonaldization: Linearity and Liquidity in Consumer Cultures," *American Behavioral Scientist* 47 (2003): 137–153.

19. Victoria de Grazia, *Irresistible Empire: America's Advance Through Twentieth-Century Europe* (Cambridge: Belknap Press of Harvard University Press, 2006).

20. Gary S. Becker and Kevin Murphy, "A Theory of Rational Addiction," *Journal of Political Economy* 96 (1988): 675–700.

21. Michael Lemonick and Alice Park, "The Science of Addiction," *Time,* July 16, 2007, 44.

22. A recent book that attempts to explore some of the patterns of irrational human behavior in this regard is Dan Ariely, *Predictably Irrational: The Hidden Forces That Shape Our Decisions* (New York: HarperCollins, 2008). A more balanced antidote to this monograph published around the same time is Tim Harford, *The Logic of Life: The Rational Economics of an Irrational World* (New York: Random House, 2008).

23. Andrew Szasz, *Shopping Our Way to Safety: How We Changed from Protecting the Environment to Protecting Ourselves* (Minneapolis: University of Minnesota Press, 2007).

24. Images of Jordan's art are viewable online at his Web site, www.chrisjordan.com.

25. "The Nylon Wars" was later published in an anthology, David Riesman, *Abundance for What? And Other Essays* (Garden City, N.J.: Doubleday, 1961). For a historical review of actual U.S. consumerist propaganda efforts, see Greg Castillo, "Domesticating the Cold War: Household Consumption as Propaganda in Marshall Plan Germany," *Journal of Contemporary History* 40 (2005): 261–288.

26. The usage of the term is also traced to the words of Joseph Goebbels, the Nazi minister of propaganda, who made a speech in 1936 in which he stated: "We can do without butter, but, despite all our love of peace, not without arms. One cannot shoot with butter, but with guns."

27. For a detailed history of this conflict see William F. Sater, *Andean Tragedy: Fighting the War of the Pacific, 1879–1884* (Lincoln: University of Nebraska Press, 2007). Also refer to Bruce W. Farcau, *The Ten Cents War: Chile, Peru, and Bolivia in the War of the Pacific, 1879–1884* (New York: Praeger, 2000).

28. For an excellent exposition of the importance of the Haber-Bosch process in the development of modern society, see Vaclav Smil, *Enriching the Earth: Fritz Haber, Carl Bosch, and the Transformation of World Food Production* (Cambridge: MIT Press, 2004).

29. David Riesman, Nathan Glazer, and Reuel Denney, *The Lonely Crowd: A Study of the Changing American Character* (New Haven: Yale University Press, 1950; rev. ed., 2001); the cover story in *Time* about Riesman's work was published on September 27, 1954; Charles McGrath, "The Big Thinkster," *New York Times Magazine,* December 29, 2002; Robert D. Putnam, *Bowling Alone: The Collapse and Revival of American Community* (New York: Simon & Schuster, 2001).

30. From the poem "The Skylark" by Percy Bysshe Shelley. For an exposition of the humanities perspective on the importance of sadness, Eric G. Wilson, *Against Happiness: In Praise of Melancholy* (New York: Farrar, Straus and Giroux, 2008), and Allan V. Horwitz and Jerome C. Wakefield, *The Loss of Sadness: How Psychiatry Transformed Normal Sorrow into Depressive Disorder* (New York: Oxford University Press, 2007). However, these books largely consider sadness as a transitional emotion, given extant human conditions, and are written in literary tradition without much empirical evidence. For a critique of this approach to the study of emotions, see Stefan Klein, *The Science of Happiness: How Our Brains Make Us Happy—and What We Can Do to Get Happier* (New York: Da Capo, 2006).

31. James B. Twitchell, *Lead Us into Temptation* (New York: Columbia University Press, 2000).

32. Quoted in Daniel Horowitz, *Anxieties of Affluence: Critques of American Consumer Culture, 1939–1979* (Boston: University of Massachusetts Press, 2005), 255.

33. Daniel Miller, "The Poverty of Morality," *Journal of Consumer Culture* 1, no. 2 (November 1, 2001): 225–243.

34. Elizabeth W. Dunn, Lara B. Aknin, and Michael I. Norton. "Spending Money on Others Promotes Happiness," *Science* 319 (2008): 1687–1688.

35. Quoted in Herbert Gintis et al., *Moral Sentiments and Material Interests: The Foundations of Cooperation in Economic Life* (Cambridge: MIT Press, 2006), 3.

36. For a general review of this field, see Marc Hauser, *Moral Minds: How Nature Designed Our Universal Sense of Right and Wrong* (New York: Ecco, 2006). The neuroeconomist's perspective is presented in Paul J. Zak, *Moral Markets: The Critical Role of Values in the Economy* (Princeton: Princeton University Press, 2008). A more eclectic perspective is presented in Michael Shermer, *The Mind of the Market: Compassionate Apes, Competitive Humans, and Other Tales from Evolutionary Economics* (New York: Times, 2007).

37. Sonja Lyubomirsky, *The How of Happiness: A Scientific Approach to Getting the Life You Want* (Penguin, 2007).

38. Daniel Kahneman, Alan B. Krueger, David Schkade, Norbert Schwarz, and Arthur A. Stone, "Would You Be Happier if You Were Richer? A Focusing Illusion," *Science* 312, no. 5782 (June 30, 2006): 1908–1910. The citation for Kahneman's Nobel Prize stated that it was awarded for his having "integrated insights from psychological research into economic science, especially concerning human judgement and decision-making under uncertainty."

39. P. Brickman and D. T. Campbell, "Hedonic Relativism and Planning the Good Society," in *Adaptation-Level Theory*, ed. Mortimer H. Appley (New York: Academic, 1971), 289.

40. Refining methods for understanding well-being is exceedingly important. As noted by Sonja Lyubomirsky in a recent interview, some of the findings of surveys are counterintuitive, and she does not feel comfortable with some of them. For example, surveys consistently show that individuals with children are less happy than those without children (across varied age groups). However, infertile couples wanting to have a child would certainly disagree with this finding, and a loss of procreation for the pursuit of well-being would clearly be detrimental for human civilization. Lyubomirsky believes that methods need to be refined to be able to tease out intervening variables that would allow for a more definitive analysis of such issues. Sonja Lyubomirsky, radio interview, *On Point*, National Public Radio, February 18, 2008, transcript available online at www.onpointradio.org.

41. In August 2005, the University of Vermont organized an interdisciplinary faculty symposium to consider the blending of objective and subjective features of human well-being, in which I was a participant. This resulted in a jointly authored paper by the faculty: R. B. Costanza, B. Fisher, S. Ali, C. Beer, L. Bond, R. Boumans, et al.,

"Quality of Life: An Approach Integrating Opportunities, Human Needs, and Subjective Well-Being," *Ecological Economics* 61 (2007): 267–276.

42. Thomas Princen, *The Logic of Sufficiency* (Cambridge: MIT Press, 2005).

43. Juliet B. Schor, *The Overspent American: Why We Want What We Don't Need* (New York: Harper Paperbacks, 1999).

44. Quoted in Benjamin Wallace Wells, "Mourning Has Broken: How Bush Privatized September 11," *Washington Monthly,* October 2003.

45. Benjamin M. Friedman, *The Moral Consequences of Economic Growth* (New York: Vintage, 2006).

46. Lewis Mumford, *Faith for Living* (New York: Harcourt Brace, 1940), 313.

Chapter 5. Curing the Resource Curse

1. Data from the United Nations Mission for the Democratic Republic of Congo (MONUC), www.monuc.org.

2. For a riveting and nuanced account of bad leadership during the Mobutu era and the impact of the Cold War on Congo, see Michela Wrong, *In the Footsteps of Mr. Kurtz: Living on the Brink of Disaster in Mobutu's Congo* (Harper Perennial, 2002). Mobutu died in exile in 1998 in Morocco but, astonishingly, his flagrant kleptocracy is still remembered by some in Congo as a time of stability. His son, Nzanga Mobutu, was elected to parliament in 2007 and made the minister of agriculture; he has his own Web site, at www.nzanga.com. The stability argument has been deconstructed by Michela Wrong (215) as follows: "Deprived of the chance to learn the lessons of its own history, Zaire's population was kept in a state of infantilism by a more insidious form of colonialism. Instead of the roller-coaster of war, destruction and eventual rebirth, the intervention of the U.S., France and Belgium, of the World Bank and the IMF, locked the society into one slow motion economic collapse. Balked of expression, unable to advance, mindsets froze over somewhere in the 1960s, leaving the country's leadership at the turn of the century stuck in an ideological time warp."

3. Estimates from the International Committee of the Red Cross, quoted in Simon Robinson and Vivienne Walt, "The Deadliest War in the World," *Time,* May 28, 2008.

4. A recent best-seller on this theme, Paul Collier, *The Bottom Billion: Why the Poorest Countries Are Failing and What Can Be Done About It* (Oxford: Oxford University Press, 2008). For an analysis of "greed versus grievance," two books published by the International Peace Academy are instructive: Karen Ballentine and Jake Sherman, eds., *The Political Economy of Armed Conflict: Beyond Greed and Grievance* (Boulder: Lynne Rienner, 2003); Mats R. Berdal and David M. Malone, eds., *Greed and Grievance: Economic Agendas in Civil Wars* (Boulder: Lynne Rienner, 2000). A much-cited volume on Venezuela's resource curse predicament is Terry Lynn Karl, *The Paradox of Plenty: Oil Booms and Petro-States* (Berkeley: University of California Press, 1997). For an excellent review of Zambia's political history, see J. B. Gewald, M. Hinfelaar, and G. Macola, *One Zambia, Many Histories: Towards a History of Post-Colonial Zambia* (Leiden: Brill, 2008). For a provocative but meticulously researched account of comparative

colonial history and postcolonial statecraft, see Jeffrey Herbst, *States and Power in Africa* (Princeton: Princeton University Press, 2000).

5. In 2002 the Belgian government officially admitted and apologized for its involvement in the assassination of Patrice Lumumba (in 1961), and established a fund of $3 million in his memory. The Belgian Commission inquiry also noted the U.S. CIA's support of the assassination. Congo's turbulent history during Belgian colonialism as well as the efforts in Europe and America to stop the cruelty of Leopold's regime is admirably documented in Adam Hochschild, *King Leopold's Ghost: A Story of Greed, Terror, and Heroism in Colonial Africa* (Boston: Mariner, 1999). Accalimed political scientist Robert Putnam's recent research on diversity concluded that there is a higher level of conflict in diverse communities. However, this may be due to the fact that we are in a time of transition between tribally charged global societies and an emergent global culture that may be more homogenized. Robert D. Putnam, "E Pluribus Unum: Diversity and Community in the Twenty-First Century, the 2006 Johan Skytte Prize Lecture," *Scandinavian Political Studies* 30, no. 2 (June 2007): 137–174.

6. The lootability hypothesis has been explored most significantly in Michael L. Ross, "A Closer Look at Oil, Diamonds, and Civil War," *Annual Review of Political Science* 9 (2006): 265–300.

7. The "Great Game" refers to the scramble for resources by the British and the Russians in Central Asia and was first used by officials in the British East India Company; the concept was introduced into English folklore by the novelist Rudyard Kipling (in his novel *Kim*). For an excellent history of this period see Peter Hopkirk, *The Great Game: The Struggle for Empire in Central Asia* (New York: Kodansha International, 1992).

8. Lord William Thompson of Kelvin, Lecture to the Institution of Civil Engineers, May 3, 1883, quoted at Today in Science History, www.todayinsci.com/K/Kelvin_Lord/Kelvin_Lord.htm (accessed June 10, 2008).

9. Richard A. Berk, *Regression Analysis: A Constructive Critique* (Thousand Oaks, Calif.: Sage, 2003). Also see Xavier Sala-i-martin, "I Just Ran Two Million Regressions," *American Economic Review* 87 (1997): 178–183.

10. William Easterly, Diego Comin, and Erick Gong, "Was the Wealth of Nations Determined in 1000 B.C.?" *Brookings Institution Global Economy and Development Working Paper* 10 (2008).

11. David A. Freedman, "Statistical Models and Shoe Leather," *Sociological Methodology* 21 (1991): 291–313.

12. Quoted in Daniel Lederman and William F. Maloney, eds., *Natural Resources, Neither Curse Nor Destiny* (Washington, D.C.: World Bank Publications, 2006), xv.

13. Mark Kurlansky, *Salt: A World History* (New York: Penguin, 2004), 225.

14. For a recent synthesis of Auty's ideas, see Richard M. Auty, ed., *Resource Abundance and Economic Development* (Oxford: Oxford University Press, 2001). Also, Jeffrey D. Sachs and Andrew M. Warner, "Natural Resources and Economic Development: The Curse of Natural Resources," *European Economic Review* 45 (2001): 827–838; a good synthesis of Collier's academic writings on this topic can be found in Paul Collier,

Breaking the Conflict Trap: Civil War and Development Policy (Washington, D.C.: World Bank Publications, 2003).

15. Paul Collier and Anke Hoeffler, "On the Economic Causes of Civil War," *Oxford Economic Papers* 50 (1998), 563–573.

16. Indra De Soysa, "Ecoviolence: Shrinking Pie or Honeypot," *Global Environmental Politics* 2 (2002): 1–36.

17. Michael Renner, *The Anatomy of Resource Wars* (Washington, D.C.: Worldwatch Institute, 2002).

18. S. Stedman, *Implementing Peace Agreements in Civil Wars: Lessons and Recommendations for Policymakers,* International Peace Academy Policy Paper, 2001, available online at www.ipacademy.org/PDF Reports/Pdf Report Implementing.pdf; M. Doyle and N. Sambanis, "International Peacebuilding: A Theoretical and Quantitative Analysis," *American Political Science Review* 94 (2000): 779–802.

19. James Fearon, Kimuli Kasara, and David Laitin, "Ethnicity, Insurgency, and Civil War," *American Political Science Review* 97, no. 1 (2003): 75–90.

20. James Fearon, "Primary Commodity Exports and Civil War," *Journal of Conflict Resolution* 49 (2005): 483–507.

21. Michael Ross, "What Do We Know About Natural Resources and Civil War," *Journal of Peace Research* 41 (2004): 337–356.

22. Macartan Humphreys, "Natural Resources, Conflict, and Conflict Resolution: Uncovering the Mechanisms," *Journal of Conflict Resolution* 49 (2005): 508–537. Also see J. Isham, M. Woolcock, L. Pritchett, and G. Busby, "The Varieties of Resource Experience: Natural Resource Export Structures and the Political Economy of Economic Growth," *World Bank Economic Review* (2006).

23. The term was first used in "The Dutch Disease," *The Economist,* November 26, 1977, 82–83. Subsequently it was analyzed in W. M. Corden, "Boom Sector and Dutch Disease Economics: Survey and Consolidation," *Oxford Economic Papers* 36 (1984): 362.

24. Graham Davis, "Learning to Love the Dutch Disease: Evidence from the Mineral Economies," *World Development* 23 (1995): 1765–1779.

25. Hossein Mahdavy, "Patterns and Problems of Economic Development in Rentier States: The Case of Iran," in *Studies in Economic History of the Middle East,* ed. M. A. Cook (London: Oxford University Press, 1970).

26. Thomas Friedman, "The First Law of Petropolitics," *Foreign Policy,* May–June, 2006.

27. Michael Ross, "Oil, Islam, and Women," *American Political Science Review* 102 (2008): 107–123.

28. Shankar Vedantam, "Oil Wealth Harms Women," *Washington Post,* March 10, 2008.

29. CNN Web site, www.cnn.com/2008/WORLD/africa/02/03/chad.explainer/.

30. Some of the more publicized works of despair regarding oil and conflict are Michael T. Klare, *Blood and Oil: The Dangers and Consequences of America's Growing Dependency on Imported Petroleum* (Holt Paperbacks, 2005); Stephen Pelletiere, *America's*

Oil Wars (Praeger, 2004); and John Ghazvinian, *Untapped: The Scramble for Africa's Oil* (Harvest, 2008).

31. Quoted in Daniel Yergin, *The Prize: The Epic Quest for Oil, Money, and Power* (New York: Free Press, 1993).

32. Mario J. Azevedo, *Roots of Violence: A History of War in Chad* (London: Routledge, 1998).

33. Jose Puppim de Oliveira and Saleem H. Ali, "New Oil in Africa: Can Corporate Behavior Transform Equatorial Guinea and Angola?" in *Corporate Citizenship in Africa,* ed. Wayne Visser et al. (Sheffield, England: Greenleaf, 2006).

34. "NYC Comptroller Wants Chevron to Review Environmental Practices," *International Herald Tribune,* December 17, 2007.

35. Quoted in Christian Nelleman, ed., *The Powers of Global Change* (Lillehammer, Norway: UNEPGRID Centre, 2003), 59.

36. For a scholarly history of Nigeria's misadventrures with oil revenues and some of the structural mistakes made in the early years of the boom, see Thomas J. Biersteker, *Multinationals, the State, and Control of the Nigerian Economy* (Princeton: Princeton University Press, 1987).

37. David Arora, *Mushrooms Demystified* (Berkeley: Ten Speed, 1986).

38. Burkhard Bilger, "The Mushroom Hunters," *New Yorker,* August 20, 2007, 62–69. Bilger also notes in his article that the world's largest organism by area may in fact be a fungal mycelium (the subterranean life form that sustains the mushrooms as terrestrial fruit) discovered in Oregon in 2003 that is believed to be more than eight thousand years old and currently covers an area of approximately two thousand acres.

39. *Japan Times,* October 26, 2007, accessed online at http://search.japantimes.co.jp/cgi-bin/nn20071026a3.html.

40. Olivier Cadot, Laurie Dutoit, and Jaime de Melo, "The Elimination of Madagascar's Vanilla Marketing Board" (Washington, D.C.: World Bank Policy Paper 3979).

41. Estimates from personal communication with various development agencies (cross-checking by author) during a field visit to Madagascar in November 2005.

42. See various reports published by the United Nations Security Council's special panel on Liberia; details at www.securitycouncilreport.org/site/c.glKWLeMTIsG/b.2400717/.

43. C. N. Brunnschweiler and E. H. Bulte, "Linking Natural Resources to Slow Growth and More Conflict," *Science* 320, no. 5876 (May 2, 2008): 616–617.

44. Macartan Humphreys, Jeffrey D. Sachs, and Joseph E. Stiglitz, *Escaping the Resource Curse* (New York: Columbia University Press, 2007).

45. Andrew Manson and Bernard Mbenga, " 'The Richest Tribe in Africa': Platinum-Mining and the Bafokeng in South Africa's North West Province, 1965–1999," *Journal of Southern African Studies* 29 (2003).

46. For a review of these various efforts at corporate social responsibility in the extractive industries in comparison to other initiatives, see Sanjeev Khagram and Saleem Ali, "Transnational Transformation: From Government-centric Interstate Regimes to Cross-sectoral Multi-level Networks of Global Governance," in *The Crisis of Global*

Environmental Governance: Towards a New Political Economy of Sustainability, ed. Jack Park, Ken Conca, and Matthias Finger (London: Routledge, 2008).

47. Erika Weinthal and Pauline Jones Luong, "Combating the Resource Curse: An Alternative Solution to Managing Mineral Wealth," *Perspectives on Politics* 4 (2006): 35–53.

48. For more details on the ways in which ownership and mineral royalty regimes have been reformed over the years, see Craig Andrews et al., *Mining Royalties: A Global Study of Their Impact on Investors, Government, and Civil Society* (Washington, D.C.: World Bank Publications, 2006).

Chapter 6. The Spoils of the Earth

1. For a thorough analysis of the spill and its aftermath, see John Keeble, *Out of the Channel: The Exxon Valdez Oil Spill in Prince William Sound,* 2nd ed. (Eastern Washington University Press, 1999).

2. Numerous academic activists have written about these movements in recent years. See, for example, Roger Moody, *Rocks and Hard Places: The Globalisation of Mining* (London: Zed, 2007), and Al Gedicks, *Resource Rebels* (Boston: South End, 2001).

3. Winona LaDuke, *All Our Relations: Native Struggles for Land and Life* (Boston: South End, 1999). A notable self-published anthology of writings by indigenous activists is Tebtebba Foundation, *Extracting Promises: Indigenous Peoples, Extractive Industries, and the World Bank* (Baguio City, Philippines: Tebtebba Foundation, 2003). For an academic analysis of the alliances between environmentalists and indigenous peoples, see Saleem H. Ali, *Mining, the Environment, and Indigenous Development Conflicts* (Tucson: University of Arizona Press, 2003).

4. People who developed certain cancers or other serious illnesses resulting from radiation exposure are eligible for fixed-amount payments under the Radiation Exposure Compensation Program if they served as uranium miners, uranium mill workers, or ore transporters, or were living downwind during atmospheric nuclear weapons tests or participated onsite during nuclear test blasts. More than half of the approximately twenty thousand claims that have beeen submitted are from downwind communities, and around forty-five hundred are from miners. On average about 70 percent of the claims are accepted for compensation under the program. Data from the U.S. Department of Justice Radiation Exposure Compensation Program, at www.usdoj.gov/civil/torts/const/reca/about.htm.

5. "Navajo Nation Pushes for Uranium Cleanup," *Morning Edition,* May 30, 2008, National Public Radio, audio transcript online at www.npr.org/templates/story/story.php?storyId=90959034. There are numerous works on the plight of the Navajo uranium miners, many with strong activist tones. The most academically rigorous account is Doug Brugge et al., *The Navajo People and Uranium Mining* (Albuquerque: University of New Mexico Press, 2007).

6. Quoted in Mark Cocker, *Rivers of Blood, Rivers of Gold: Europe's Conflict with Tribal Peoples* (London: Jonathan Cape, 1998), 288.

7. Quoted in Geoffrey Wheatcroft, *The Randlords* (Weidenfeld and Nicolson History, 1993), 64.

8. Quoted in Richard Manning, *One Round River: The Curse of Gold and the Fight for the Big Blackfoot* (New York: Henry Holt, 1998), 123.

9. Georgius Agricola, *De Re Metallica* (New York: Dover, 1950), 8, translated by Herbert Clark Hoover and his wife Lou Henry Hoover in 1913; original German version published in 1556.

10. Quoted in Rebecca Solnit, "Winged Mercury and the Golden Calf: Two Elements, One Economic Theory, and a Cascading Torrent of Collateral Damage," *Orion*, September–October 2006, 19.

11. Quoted in Duane A. Smith, *Mining America: The Industry and the Environment, 1800–1980* (Lawrence: University Press of Kansas, 1986), 47.

12. Sierra Club Canada press release, www.sierraclub.ca/national/media/inthenews/item.shtml?x=2777.

13. The classic work on the sociology of Appalachian communities in this regard is John Gaventa, *Power and Powerlessness: Quiescence and Rebellion in an Appalachian Valley* (Urbana: University of Illinois Press, 1982). See also Cynthia M. Duncan, *Worlds Apart: Why Poverty Persists in Rural America* (New Haven: Yale University Press, 2000).

14. The story of Judy Bonds and her community is told in Michael Shnayerson, *Coal River* (New York: Farrar, Straus and Giroux, 2008).

15. Stuart Kirsch, *Reverse Anthropology: Indigenous Analysis of Social and Environmental Relations in New Guinea* (Stanford: Stanford University Press, 2006). For a detailed analysis of how environmentalism has shaped governance in New Guinea as a result of such efforts, see Paige West, *Conservation Is Our Government Now: The Politics of Ecology in Papua New Guinea* (Durham: Duke University Press, 2006).

16. For an excellent analysis of the Bougainville conflict and citations to most of the relevant material, see Chris Ballard and Glenn Banks, "The Anthropology of Mining," *Annual Review of Anthropology* 32 (2003): 287–313. A company perspective on the Bougainville conflict can be found in the narrative of the mine's managing director, who lived in the region for twenty years; Paul Quodling, *Bougainville: The Mine and the People* (Auckland: Centre for Independent Studies, 1991).

17. June Nash, *We Eat the Mines and the Mines Eat Us* (New York: Columbia University Press, 1993), ix. For more recent work on Bolivian artisanal mining, see Eduardo Quiroga, "The Case of Artisanal Mining in Bolivia: Local Participator Development and Mining Investment Opportunities," *Natural Resources Forum* 26 (2002): 127–139.

18. Friedrich Engels, *The Conditions of the Working Class of England in 1844,* text online at www.marxists.org/archive/marx/works/1845/condition-working-class/ch11.htm.

19. "Coal Mine Canaries Made Redundant," BBC news site, http://news.bbc.co.uk/onthisday/hi/dates/stories/december/30/newsid_2547000/2547587.stm.

20. Quoted in Barbara Freese, *Coal: A Human History* (London: Penguin, 2002), 84.

21. For a review of gender issues related to mining, see Kuntala Lahiri-dutt, *Women Miners in Developing Countries: Pit Women and Others* (Burlington, Vt.: Ashgate, 2006).

22. Alan Gallop, *Children of the Dark: Life and Death Underground in Victoria's England* (London: Sutton, 2003). The health impacts of coal mining are an ongoing concern across the world. For a review of these concerns see M. H. Ross and J. Murray, "Occupational Respiratory Disease in Mining," *Occupational Medicine* 54 (2004): 304–310. The impacts of coal mining can range from pulmonary ailments, such as lung cancer and emphysema, to fluorosis, nephropathy, and arsenic poisoning from impurities in the coal seams. See Robert B. Finkelman et al., "Health Impacts of Coal and Coal Use: Possible Solutions," *International Journal of Coal Geology* 50 (2002): 425–443.

23. Jean-Paul Richalet et al., "Chilean Miners Commuting from Sea Level to 4500m: A Prospective Study," *High Altitude Medicine and Biology* 3, no. 2 (2002).

24. "Ghana Gold Workers Paid in Condoms," BBC online news story, February 19, 2003.

25. C. Campbell et al., "Is Social Capital a Useful Conceptual Tool for Exploring Community Level Influences on HIV Infection?" *AIDS Care* 14 (2002): 41–54.

26. In some cases the mining areas can also be clusters of drug-resistant malaria, as was found in one study in Venezuela; A. Ache et al., "In Vivo Drug Resistance of Falciparum Malaria in Mining Areas of Venezuela," *Tropical Medicine and International Health* 7 (2002): 737–743. For a detailed review of various health and safety issues related to small-scale mining, see Gavin M. Hilson, *Small-Scale Mining, Rural Subsistence, and Poverty in West Africa* (London: Practical Action, 2008).

27. F. B. Pyatt and J. P. Grattan, "Some Consequences of Ancient Mining Activities on Health of Ancient and Modern Human Populations," *Journal of Public Health Medicine* 23 (2001): 235–236.

28. Details about the Roman mining law were found in the German Mining Museum, Bochum, Germany, which has some of the earliest Roman tablets recording the history of mining in the Iberian peninsula (visited in May 2008).

29. Richard J. Davies et al., "The East Java Mud Volcano: An Earthquake or Drilling Trigger?" *Earth and Planetary Science Letters,* published online, June 5, 2008.

30. These incidents are documented in the *Congressional Record* and covered in several news articles. The *Charleston Gazette* has a Web compendium of articles and incident reports pertaining to the Sago tragedy, at http://wvgazette.com/News/The+Sago+Mine+Disaster.

31. Juan Pablo Pérez Alfonzo was one of the founders of OPEC and died in 1979. There are innumerable references to this quotation, although I have been unable to find an exact citation for when he actually said it.

32. Roy Bates, *Chinese Dragons* (New York: Oxford University Press, 2002).

33. Judith Shapiro, *Mao's War Against Nature: Politics and the Environment in Revolutionary China* (New York: Cambridge University Press, 2001).

34. Pig iron is the first intermediary product extracted from raw ore through heating in a blast furnace fired with coke. In early production, the molding of the ingots

along a central runner resembled suckling piglets and gave the product its unlikely name.

Most of the iron China produces is believed to be mined in China, but figures on actual output from Chinese mines versus scrap usage and iron imports are not delineated. China is also the world's largest scrap importer, with more than 5 million tons annually, worth nearly $5 billion in 2006.

35. "Angling for Iron Ore in China's Streams," Reuters News Service, October 23, 2007.

36. This estimate was based on annual data from 2005; average statistics across numerous years would be a preferable comparison; Michael Moser, "Coal Mine Safety in the U.S. and China," presentation at Resources for the Future, Washington, D.C., February 1, 2006.

37. Indrajit Basu, "China Boosts Domestic Mining," *UPI Asia Online*, May 29, 2008, http://upiasiaonline.com/Economics/2008/05/29/china_boosts_domestic _mining/6152/.

38. The project details are available at UNDP's China Web site, www.undp.org.cn/.

39. "China Coal Mining Fatalities Drop 20 Percent in 2007," *Forbes*, January 13, 2008, www.forbes.com/afxnewslimited/feeds/afx/2008/01/13/afx4524639.html.

40. R. Q. Li, M. Dong, Y. Zhao, L. L. Zhang, Q. G. Cui, and W. M. He, "Assessment of Water Quality and Identification of Pollution Sources of Plateau Lakes in Yunnan (China)," *Journal of Environmental Quality* 36 (2007): 291–297.

41. D. G. Streets et al., "Anthropogenic Mercury Emissions in China," *Atmospheric Environment* 39 (2005): 7789–7806.

42. "Nine Die as China House Collapses in Cyanide," Reuters News service, September 28, 2007.

43. Cyanide incidences are tracked by the Mineral Policy Institute in Australia, www .mpi.org.au/campaigns/cyanide/cyanide_spills/.

44. For guidelines on closure planning from Australia, which has learned from its past mistakes considerably, see David Laurence, "Optimising Mine Closure Outcomes for the Community—Lessons Learnt," *Minerals and Energy* 17 (2002): 27–35.

45. Gui-Bin Jiang, Jian-Bo Shi, and Xin-Bin Feng, "Mercury Pollution in China," *Environmental Science and Technology* 40 (2006): 3672–3678.

46. United States Geological Survey, *China Minerals Outlook* (Washington, D.C.: USGS Publications).

47. Indrajit Basu, "China Boosts Domestic Mining," *UPI Asia Online*, May 29, 2008, http://upiasiaonline.com/Economics/2008/05/29/china_boosts_domestic_mining/6152/.

48. "China Bans Mining on Sacred Buddhist Mountain," Reuters News Service, August 24, 2007.

49. A. J. Gunson, *Mercury and Artisanal and Small-Scale Gold Miners in China* (Master's thesis, University of British Columbia, Department of Mining Engineering, 2005).

50. See the Web site of the program in Chinese, www.casmchina.org.

51. Partha Dasgupta, *Human Well-Being and the Natural Environment* (New York: Oxford University Press, 2004), 109.

52. For a highly readable account of Nauru, as well as some of the adjoining islands that were able to avert such resource crashes, such as Tikopia, see Carl N. McDaniel and John M. Gowdy, *Paradise for Sale: A Parable of Nature* (Berkeley: University of California Press, 2000).

53. Mark Sagoff, *The Economy of the Earth: Philosophy, Law, and the Environment* (New York: Cambridge University Press, 2007).

54. "Payments for Ecosystem Services" conference, organized by the Gund Institute for Ecological Economics and the Moore Foundation, held in San Jose, Costa Rica, March 8–16, 2007. Remarks recorded by author in attendance at the conference.

55. Speech at the State of the World Conference, Aspen, Colorado, July 22, 2001, quoted in Lester R. Brown, *Plan B: Mobilizing to Save Civilization* (New York: W. W. Norton, 2007).

Chapter 7. Destination Cradle

1. Data on salt consumption from the Salt Institute, www.saltinstitute.org.

2. For a good review of the coltan sector's contribution to conflict but also how the industry can be positively transformed, see K. Hayes and R. Burge, *Coltan Mining in the Democratic Republic of Congo: How Tantalum-Using Industries Can Commit to the Reconstruction of the DRC* (London: Fauna and Flora International, 2003). Miners who work in these jungle mines frequently hunt down endangered species, such as gorillas, for bush meat, which is an additional environmental concern of this enterprise.

3. R. Abad, "Squatting and Scavenging in Smokey Mountain," *Philippine Studies* 39 (1991): 267–285.

4. H. Spooner, *Wealth from Waste: Elimination of Waste a World Problem* (London: George Routledge and Sons, 1918).

5. Martin Medina, *The World's Scavengers: Salvaging for Sustainable Consumption and Production* (New York: Altamira, 2007).

6. For an account of how the hazardous-waste sector was influenced by the Mafia and the serious environmental consequences of its control in many urban areas during the 1970s and 1980s, see Alan A. Block and Frank R. Scarpitti, *Poisoning for Profit: The Mafia and Toxic Waste in America* (New York: William Morrow, 1985).

7. Robin Pomeroy, "Naples Garbage Is Mafia Gold," Reuters News Service article, January 9, 2008.

8. John Seabrook, "American Scrap: An Old-School Industry Globalizes," *New Yorker*, January 14, 2008.

9. Steel recycling and sustainability information from the Steel Recycling Institute's environmental Web site, www.sustainable-steel.org.

10. For an academic analysis of the 1979 film and its remarkable influence (holding a record for the most profitable ratio of any movie until 2000, when it was overtaken by *The Blair Witch Project*), see Mick Broderick, "Heroic Apocalypse: Mad Max,

Mythology, and the Millennium," in *Crisis Cinema: The Apocalyptic Idea in Postmodern Narrative Film,* ed. Christopher Sharrett (New York: Maisonneuve, 1993).

11. News release on the Chinese government Web site, January 24, 2002, www1.10thnpc.org.cn/english/2002/Jan/25776.htm.

12. A philosophical exposition of "the creation and destruction of value" is explored in Michael Thompson, *Rubbish Theory* (Oxford: Oxford University Press, 1979). Thompson begins with an amusing riddle: "What does a poor person throw away that a rich man puts in his pocket?" The answer is "Snot!" Social ethos has made disposal of human waste more circumscribed for the wealthy. A more recent philosophical excursion into garbage is Greg Kennedy, *An Ontology of Trash: The Disposable and Its Problematic Nature* (Albany: SUNY Press, 2007).

13. William Rathje and Cullen Murphy, *Rubbish! The Archaeology of Garbage* (New York: HarperCollins, 1993; 2nd ed., Tucson: University of Arizona Press, 2001).

14. This term was first used by *Business Week* in the context of consumer goods in 1955; quoted in Susan Strasser, *Waste and Want: A Social History of Trash* (New York: Holt Paperbacks, 2000), 274.

15. United States Environmental Protection Agency, *Municipal Solid Waste Generation, Recycling, and Disposal in the United States* (Washington, D.C.: EPA, 2006).

16. Frank B. Golley, *A History of the Ecosystem Concept in Ecology: More than the Sum of the Parts* (New Haven: Yale University Press, 1996), 8.

17. Norbert Wiener, *Cybernetics: Or the Control and Communication in the Animal and the Machine* (Cambridge: MIT Press, 1948). The ecological applications of the field are discussed in A. M. Andrews, "Ecofeedback and Significance Feedback in Neural Nets and in Society," *Journal of Cybernetics* 4 (1974): 61–72.

18. Thomas E. Graedel and Braden R. Allenby, *Industrial Ecology* (Englewood Cliffs, N.J.: Prentice Hall, 2002). Since the publication of their book both Graedel and Allenby have moved to academic career paths at universities. Another important work in the same genre but from the perspective of social science and product architecture is William McDonough and Michael Braungart, *Cradle to Cradle: Remaking the Way We Make Things* (New York: North Point, 2002).

19. R. U. Ayres and U. E. Simonis, eds., *Industrial Metabolism* (Tokyo: United Nations University Press, 1994).

20. C. Piluso, Y. Huang, and H. H. Lou, "Ecological Input-Output Analysis-Based Sustainability Analysis of Industrial Systems," *Industrial and Engineering Chemistry Research* 47 (2008): 1955–1966.

21. Ibrahim Dincer and Marc A. Rosen, *Exergy: Energy, Environment, and Sustainable Development* (Amsterdam: Elsevier Science, 2007).

22. Details of the program are at www.nisp.org.uk. For a comparative analysis of U.S. and European efforts in this regard, see R. R. Heeres and W. J. V. Vermeulen, "Eco-Industrial Park Initiatives in the USA and the Netherlands: First Lessons," *Journal of Cleaner Production* 12 (2004): 985–995. The U.S. Environmental Protection Agency is also trying to facilitate eco-industrial park development, though the effort is

somewhat scattered at present; details of projects and related links are at www2.ucsc.edu/gei/eco-industrial_parks.html.

23. J. Weiss, "Sweden Solving the Four E's: Economics, Employment, Environment, Energy," *Europe* 349 (1995): 6–9.

24. Dara O'Rourke, "Industrial Ecology: A Critical Review," *International Journal of Environment and Pollution* 6 (1996): 89–112.

25. An example of such techniques is the concept of Mediated Modeling, introduced in Marjan van den Belt, *Mediated Modeling: A System Dynamics Approach to Environmental Consensus Building* (Washington, D.C.: Island, 2004).

26. Robert M. Solow, "The Economics of Resources or the Resources of Economics," *American Economic Review* 64, no. 2 (1974): 1–14.

27. V. Brodyansky et al., *The Efficiency of Industrial Processes: Exergy Analysis and Optimization* (Amsterdam: Elsevier, 1994).

28. The award itself is quite literally a treasure since it is based on the emperor's three jewels: the Yata mirror, the Yasakani jewel (a precious jade), and the emperor's personal sword. There are eight "classes" of the order, in ascending numerals of importance, and Deming was awarded the second class.

29. The Society for Environmental Toxicology and Chemistry (SETAC) has some clear guidelines on LCA, at www.setac.org/node/32.

30. Chris Hendrickson, Lester Lave, and H. Scott Matthews, *Environmental Life Cycle Assessment of Goods and Services: An Input-Output Approach* (Washington, D.C.: Resources for the Future, 2006).

31. Paul Shrivastava, "Ecocentric Management in Industrial Ecosystems: Management Paradigms for a Risk Society," *Academy of Management Review* 20 (1995): 118–127.

32. For an excellent review of data on dematerialization from 1980 to 2006 and some policy indicators on materials management, see Jesse H. Ausubel and Paul E. Waggoner, "Dematerialization: Variety, Caution, and Persistence," *Proceedings of the National Academy of Sciences* 105 (35): 12774–12779, September 2008.

Chapter 8. The Restoration Ethic

1. There are several studies of this project, including Steve Bogener, *Ditches Across the Desert: Irrigation in the Lower Pecos Valley* (Lubbock: Texas Tech University Press, 2003); Donald J. Pisani, *Water and American Government: The Reclamation Bureau, National Water Policy, and the West, 1902–1935* (Berkeley: University of California Press, 2002); Mark Hufstetler and Lon Johnson, *Watering the Land: The Turbulent History of the Carlsbad Irrigation District* (Denver: National Park Service, Rocky Mountain Region, Division of National Preservation Programs, 1993).

2. Superfund is the common name for the Comprehensive Environmental Responsibility Compensation and Liability Act, passed by the U.S. Congress in 1980. Although there was some funding allocated for cleanup of Superfund sites, it is woefully inadequate for the full expenses required, and the aim is to find "Potentially Responsible Parties" to pay for the cleanup. Much of the $6 billion fund was exhausted in 2003,

and where no responsible parties can be found to pay for current cleanups, the money must be authorized by Congress and allocated from general funds.

3. Details of Janice Valera's involvement in this case are derived from Trout Unlimited, "The Grassroots Guide to Abandoned Mine Cleanup" (Arlington, Va., 2005).

4. Details of the technical aspects of the restoration effort can be found in Paul Robinson, "Innovative Administrative, Technical, and Public Involvement Approaches to Environmental Restoration at an Inactive Lead-Zinc Mining and Milling Complex near Pecos, New Mexico," in *Proceedings of Waste Management '95, University of Arizona/DOE/WEC* (Tucson, March 1995).

5. Richard Manning, *One Round River: The Curse of Gold and the Fight for the Big Blackfoot* (New York: Henry Holt, 1998), 146.

6. Karl Gustavson et al., "Superfund and Mining Megasites," *Environmental Science and Technology* 41 (2007): 2667–2672.

7. Jim Kuipers, Center for Science in Public Participation, "Putting a Price on Pollution," Mineral Policy Center Issue Paper no. 4 (2003).

8. Details about the Giant Mine reclamation effort can be found at a special Web site created by the Canadian government to inform the public of the procedure to immobilize the toxins, at http://nwt-tno.inac-ainc.gc.ca/giant/.

9. Details on land ownership in Scotland and the history of laird-tenant relations can be found in J. McEwen, *Who Owns Scotland: A Study in Land Ownership* (Edinburgh: Edinburgh University Student Publication Board); A. Wightman and P. Higgins, "Sporting Estates and the Recreational Economy in the Highlands and Islands of Scotland," *Scottish Affairs* 31 (2000): 18–36.

10. The idea is also credited to Colin Gribble, a geologist at the University of Glasgow, and Ian Wilson, a Scottish entrepreneur specializing in minerals to supply aggregate for a third London airport; R. Cowell, "Localities and the International Trade in Aggregates: Coastal Superquarries on the Horizon?" *Geography* 85 (2000): 134–144.

11. Gold and silver mineral rights are governed by the Royal Mines Act of 1424, petroleum and natural gas by the Petroleum Production Act of 1934; coal was nationalized in 1947 and is now managed by the Coal Authority.

12. Details about the Glensanda quarry based on a field visit by the author to the site and interviews with site management.

13. Alastair McIntosh, personal communication, Edinburgh, June 2, 2004. In a letter he wrote to Lord Maxwell McLeod, a proponent of Glensanda, McIntosh stated, "I accept that quarrying is necessary and that Glensanda does it in a relatively right-on way"; letter dated April 9, 2003. See also Alastair McIntosh, *Soil and Soul: People Versus Corporate Power* (London: Aurum, 2004).

14. The wind-farm development is controversial for aesthetic reasons despite its environmental advantages. Some opposition groups are advocating against the wind-turbine project; www.fairwind.org.uk.

15. E. O. Wilson, *The Diversity of Life* (Cambridge: Harvard University Press, 1992), 269.

16. Conservation International has listed New Caledonia as one of its "biodiversity hotspots"; Michel Pascal et al., "Mining and Other Threats to the New Caledonia Hotspot," *Conservation Biology* 22 (2007): 498–499.

17. For a detailed study of the investment of two mining companies in New Caledonia, see Saleem H. Ali and Andrew Singh Grewal, "The Ecology and Economy of Indigenous Resistance," *Contemporary Pacific* 18 (2006): 361–392.

18. Details of the Cornwall rehabilitation efforts are from an editorial in *Mining Environmental Management Magazine* (January 2004).

19. A competitive ecological challenge prize in Fuller's honor, worth $100,000, was inaugurated in 2008, and the first recipient was applied biologist John Todd, whose proposal for remediating coal-mining lands was commended by the jury for "weaving together a set of processes—from restoration of land to geo-sequestration of carbon, to community involvement, to long-term economic vitality—to create a blueprint for a future for Appalachia that envisions a harmonious self-sustaining community." Details at www.bfi.org. For an account of the life and times of Buckminster Fuller, see Lloyd Steven Sieden, *Buckminster Fuller's Universe: His Life and Work* (New York: Basic, 2000).

20. A good exposition of such plants can be found in an anonymous editorial titled "Mining, Metallophytes, and Land Reclamation," *Mining Environmental Management Magazine* 10 (March 2002): 11–16.

21. Details about the center can be found at www.eica-ratho.com/content/about-us/1120/.

22. D. N. Pandey et al., "Mine Spoil Restoration: A Strategy for Combining Rainwater Harvesting and Adaptation to Random Recurrence of Droughts in Rajasthan," *International Forestry Review* 7 (2005): 241–249.

23. An intriguing history of the prison and its role in nineteenth-century American history is chronicled in Richard Phelps, *A History of Newgate of Connecticut* (Albany: J. Musell, 1860); manuscript available online through the digitized books project of Google Books, at http://books.google.com.

24. See James B. McGraw, "Evidence for Decline in Stature of American Ginseng Plants from Herbarium Specimens," *Biological Conservation* 98 (2001): 25–32.

25. The impairment of Appalachia's waterways has been studied in detail. One recent study that uses bioassessment indices to measure impacts is Jason Freund and J. Todd Petty, "Response of Fish and Macroivertebrate Bioassessment Indices to Water Chemistry in a Mined Appalachian Watershed," *Environmental Management* 39 (2007): 707–720.

26. Details recorded by author in a visit to the Melbourne Museum, Australia, July 2007.

27. Best Practice Profile of the Argyle Diamond Mine, Rio Tinto Corporation, December 15, 2006, document obtained by author during a site visit to the Argyle mine in July 2007.

28. Human Rights and Equal Opportunity Commission of Australia, Native Title Report (Canberra: Government Printing Office, 2006).

29. James Otto et al., *Mining Royalties: A Global Study of Their Impact on Investors, Government, and Civil Society* (Washington, D.C.: World Bank, 2007), 147.

30. Details of the cost of cleanup and the security bond obtained from Mining Watch Canada, March 2008.

31. The term "race to the bottom" was coined by Justice Louis Brandeis of the U.S. Supreme Court in *Ligget County v. Lee* (1933) 288 U.S. 517, 558–559. For more details about this concept and its manifestations in corporate behavior, see Alan Tonelson, *The Race to the Bottom: Why a Worldwide Worker Surplus and Uncontrolled Free Trade Are Sinking American Living Standards* (New York: Basic, 2002).

32. Details about Bisbee's economic recovery are from Cochise College Center for Economic Research, *Bisbee Outlook* (Bisbee, Ariz.: City of Bisbee, 2006).

33. See the research products ensuing from this project, conducted by the author, including a documentary video for viewing online at www.fijigold.org.

34. Cecily Neil et al., eds., *Coping with Closure: An International Comparison of Mine Town Experiences* (London: Routledge, 1992), 22.

35. For an excellent review of controversies in the field, see Andre F. Clewell and James Aronson, *Ecological Restoration: Principles, Values, and Structure of an Emerging Profession* (Washington, D.C.: Island, 2008).

36. Jack E. Williams, Michael P. Dombeck, and Christopher A. Wood, *From Conquest to Conservation: Our Public Lands Legacy* (Washington, D.C.: Island, 2003).

Chapter 9. Embracing the Treasure Impulse

1. The initial use of the term "Scotch" was meant to be a caustic joke by the marketing arm of the company referring to 3M's founders, who are of Scottish origin. Initially they had resisted putting adhesive on the full tape, and a local businessman referred to them as "stingy Scotch." The name clung on quite effectively (historical references to this etymology can be found in the 3M Museum, Twin Harbors, Minnesota).

2. The company has won many awards for its efforts from the U.S. EPA and also a listing as a "cool company" on the Web site www.cool-companies.org.

3. A description of this product development case is found in Daniel C. Esty and Andrew S. Winston, *Green to Gold: How Smart Companies Use Environmental Strategy to Innovate* (New Haven: Yale University Press, 2006).

4. Description based on an interview with Agarwal aired on CNBC-India's program *Khas Mulakaat* in Hindi on March 28, 2008.

5. Georgius Agricola, *De Re Metallica* (Dover, 1950), translated by Herbert Clark Hoover and his wife Lou Henry Hoover in 1913; original German version published in 1556.

6. Andrew Carnegie, "Mr. Carnegie's Address," in Presentation of the Carnegie Library to the People of Pittsburgh, with a Description of the Dedicatory Exercises, November 5, 1895 (Pittsburgh: City of Pittsburgh, [1895]), 13–14; from the Web site of the Carnegie Library of Pittsburgh, www.clpgh.org/exhibit/carnegie.html (accessed April 4, 2008).

7. For a discussion of the power the nineteenth-century tycoons had in shaping the U.S. economy, see Charles R. Morris, *The Tycoons: How Andrew Carnegie, John D. Rockefeller, Jay Gould, and J. P. Morgan Invented the American Supereconomy* (New York: Holt Paperbacks, 2006).

8. See Kenneth W. Rose and Darwin H. Stapleton, "Toward a 'Universal Heritage': Education and the Development of Rockefeller Philanthropy, 1884–1913," *Teachers College Record* 93, no. 3 (1992): 536–555.

9. Compiled from the Families of Philanthropy Web site, www.familiesofphilanthropy.com/rockefellertl.html (accessed April 3, 2008).

10. For an excellent history of these men of fortune in Africa who thrived on the colonial enterprise, see Geoffrey Wheatcroft, *The Randlords* (London: Weidenfeld and Nicolson History, 1993).

11. Susan U. Raymond, *The Future of Philanthropy: Economics, Ethics, and Management* (Hoboken, N.J.: Wiley, 2004).

12. Allison N. McCoy and Michael L. Platt, "Risk Sensitive Neurons in Macaque Posterior Cingulate Cortex," *Nature Neuroscience* 8 (2005): 1220–1227. For a monumental history of gambling, see David G. Schwartz, *Roll the Bones: The History of Gambling* (New York: Gotham, 2007).

13. For an engaging account of the philosophical aspects of *Star Trek*, see Judith Barad and Ed Robertson, *The Ethics of Star Trek* (New York: Harper Perennial, 2001). On the scientific validity of some *Star Trek* narratives and the rejection of others, see Lawrence Krauss, *The Physics of Star Trek*, rev. ed. (New York: Basic, 2007).

14. For a documentation of conflicts over water see the database prepared by Aaron Wolf and colleagues at Oregon State University, at www.transboundarywaters.orst.edu/.

15. "Taliban in Texas for Talks on Gas Pipeline," BBC News, December 4, 1997, http://news.bbc.co.uk/2/hi/world/west_asia/37021.stm. For a discussion of the potential for oil and peace building, see Jill Shankleman, *Oil, Profits, and Peace: Does Business Have a Role in Peacemaking?* (Washington, D.C.: United States Institute of Peace Press, 2007).

16. The term "cornucopia" is derived from Greek and Roman mythology, and means "horn of abundance." Cassandra was a prophetess of Troy, whose dire predictions were often not believed by the public, even if they might be true.

17. Ira Sohn, "Long-Term Projections of Non-Fuel Minerals: We Were Wrong but Why?" *Resources Policy* 30 (2006): 259–284.

18. Thomas E. Graedel and Robert J. Klee, "Getting Serious About Sustainability," *Environmental Science and Technology* 36 (2002): 523–529.

19. Joshua Davis, "Race to the Bottom," *Wired,* March 2007.

20. Quoted in the *Johannesburg Sunday Business Times,* June 6, 2004, www.sundaytimes.co.za/2004/06/06/business/news/news18.asp.

21. These scenario descriptions are from MMSD North America, *Learning from the Future* (Winnipeg: International Institute for Sustainable Development, 2002).

22. There has been much controversy over this law and reform bills have been deliberated for decades in Congress. In 2008, the House of Representatives passed the Hard Rock Mining and Reclamation Act, but final enactment of the law is still uncertain. For updates on mining reform, see the Web site set up by Earthworks, www.earthworksaction.org/us_program.cfm. For an excellent exposition of the legal history of hard-rock mining in the United States, see John F. Seymour, "Hardrock Mining and the Environment: Issues of Federal Enforcement and Liability," *Ecology Law Quarterly* 31 (2004): 795–956, and Bart Lounsbery, "Digging Out of the Holes We Have Made: Hardrock Mining, Good Samaritans, and the Need for Comprehensive Action," *Harvard Environmental Law Review* 32 (2008): 149–216. Also see Tyler Weidlich, "The Mining Law Continuum: Is There a Contemporary Prospect for Reform?" *Brandeis Law Journal* 44 (2006): 951–977.

23. This scenario is reminiscent in some ways of what is proposed in James Howard Kunstler, *The Long Emergency: Surviving the End of Oil, Climate Change, and Other Converging Catastrophes of the Twenty-First Century* (New York: Atlantic Monthly, 2005).

24. The proceedings of this workshop and the structural analysis the community conducted to help in decision making are documented in International Institute for Sustainable Development, *Out of Respect: The Tahltan, Mining, and the Seven Questions to Sustainability* (Winnipeg: IISD, 2003).

25. For an international development analysis of the notion of "free prior and informed consent," see United Nations Department of Economic and Social Affairs, "Free Prior and Informed Consent and Beyond" (New York: United Nations Document PFII/2005/WS.2/10).

26. The classic paper on the topic is Charles E. Lindblom, "The Science of 'Muddling Through,'" *Public Administration Review* 19, no. 2 (Spring 1959): 79–88. Also Charles E. Lindblom, "Still Muddling, Not Yet Through," *Public Administration Review* 39, no. 6 (December 1979): 517–526.

27. Herman E. Daly, *Beyond Growth: The Economics of Sustainable Development* (Boston: Beacon, 1997).

28. Benjamin M. Friedman, *The Moral Consequences of Economic Growth* (New York: Vintage, 2006).

29. William Easterly, *The Elusive Quest for Growth: Economists' Adventures and Misadventures in the Tropics* (Cambridge: MIT Press, 2002); William Easterly, *The White Man's Burden: Why the West's Efforts to Aid the Rest Have Done So Much Ill and So Little Good* (New York: Penguin, 2007).

30. Bill McKibben, *Deep Economy: The Wealth of Communities and the Durable Future* (New York: Holt Paperbacks, 2008).

31. Named after the Russian-American economist and Nobel Prize winner Simon Kuznets. For an academic review of his life and work, see Robert W. Fogel, "Simon S. Kuznets: April 30, 1901–July 9, 1985," NBER Working Paper No. W7787 (2000).

32. David I. Stern, "The Rise and Fall of the Environmental Kuznets Curve," *World Development* 32 (2004): 1419–1439. For a study of how the environmental Kuznets curve explains some kinds of pollution, such as air and noise, but not other forms of

environmental harm, see Matthew E. Kahn, *Green Cities: Urban Growth and the Environment* (Washington, D.C.: Brookings Institution Press, 2006).

33. Study cited by Brice Pedroletti, "En Chine, le déficit de politique écologique menace les performances économiques," *Le Monde,* July 2, 2005. This point is expanded in Saleem H. Ali, "In China Globalization Can Be Green," *International Herald Tribune,* May 30, 2006.

34. The research by Bowles was part of a multiyear McArthur Foundation project that resulted in notable books in collaboration with the Russell Sage Foundation: Samuel Bowles, Steven N. Durlauf, and Karla Hoff, *Poverty Traps* (Princeton: Princeton University Press, 2006); Jean-Marie Baland, Pranab Bardhan, and Samuel Bowles, *Inequality, Cooperation, and Environmental Sustainability* (Princeton: Princeton University Press, 2006). In addition, research by James Boyce included in *Inequality, Cooperation, and Environmental Sustainability* has linked inequality and environmental decline, perhaps closing the circle on the need for pricing nature at some level. Research at the Global Development and Environment Institute at Tufts University is also opening avenues for communication, through engagement on such matters as environmental reform proposals on world trade.

35. I recall conferring with the economist Robert Mendelsohn during my graduate school days about the work of E. F. Schumacher, who wrote the book *Small Is Beautiful* (1973). Mendelsohn immediately dismissed him as a "neo-Marxist," even though most Marxist economics was about large-scale industrialization and showed little care for the environment, which was quite antithetical to Schumacher's views. The only commonality was perhaps concern about inequality in wealth distribution, which is now being recognized as a problem by even mainstream economists. For an excellent intellectual history of economics, see Eric D. Beinhocker, *Origin of Wealth: Evolution, Complexity, and the Radical Remaking of Economics* (Boston: Harvard Business School Press, 2007).

36. Joseph E. Stiglitz, *Globalization and Its Discontents* (New York: W. W. Norton, 2003). Accusations by Stiglitz about the International Monetary Fund drew an acrimonious response in Kenneth Rogoff, "The IMF Strikes Back: An Open Letter to Joe Stiglitz," *Foreign Policy* 134 (January–February 2003). Stiglitz was a bit more tempered in his next book, Joseph E. Stiglitz, *Making Globalization Work* (New York: W. W. Norton, 2007).

37. Sebastian Mallaby, *The World's Banker: A Story of Failed States, Financial Crises, and the Wealth and Poverty of Nations* (New York: Penguin, 2006).

38. Roy Carr-Hill and John Lintott, *Consumption, Jobs, and the Environment: A Fourth Way?* (New York: Palgrave Macmillan, 2003).

39. Richard H. Thaler and Cass R. Sunstein, *Nudge: Improving Decisions About Health, Wealth, and Happiness* (New Haven: Yale University Press, 2008).

40. David Vogel, *The Market for Virtue: The Potential and Limits of Corporate Social Responsibility* (Washington, D.C.: Brookings Institution Press, 2006), 3.

41. Newt Gingrich and Terry Maple, *A Contract with the Earth* (Baltimore: Johns Hopkins University Press, 2008).

42. Non-Han Chinese are allowed to have more than one child. Even among the Han Chinese population the one-child policy is frequently violated. Forced abortions have occasionally been recorded in certain jurisdictions, although they are not allowed by law.

43. Garrett Hardin was the author of the most cited paper in *Science* magazine, "The Tragedy of the Commons." Later in life he became a staunch advocate of draconian population control (in spite of having five children himself). His later works on the topic include Garrett Hardin, "Lifeboat Ethics: The Case Against Helping the Poor," *Psychology Today,* September 1974.

44. Amartya Sen, "The Impossibility of a Paretian Liberal," *Journal of Political Economy* 78, no. 1 (January–February 1970): 152–157. Replies to this article included Y. K. Ng, "The Possibility of a Paretian Liberal: Impossibility Theorems and Cardinal Utility," *Journal of Political Economy* 79, no. 6 (November–December 1971): 1397–1402; Claude Hillinger and Victoria Lapham, "The Impossibility of a Paretian Liberal: Comment by Two Who Are Unreconstructed," *Journal of Political Economy* 79, no. 6 (December 1971): 1403–1405. Sen responded to these critiques in Amartya Sen, "The Impossibility of a Paretian Liberal: Reply," *Journal of Political Economy* 79, no. 6 (December 1971): 1406–1407.

45. Kenneth J. Arrow, *Social Choice and Individual Values,* 2nd ed. (New Haven: Yale University Press, 1970).

46. This analogy was given first in Julian Simon, *The Ultimate Resource 2* (Princeton: Princeton University Press, 1998). The analogy may be specious in the context of finite usable extractive resources on earth, but it is quite appropriate for considering human ingenuity within the constraints of natural laws. For an anthropological analysis of how circular thinking is pervasive in human societies, see Mary Douglas, *Thinking in Circles: An Essay on Ring Composition* (New Haven: Yale University Press, 2007).

47. Aubrey de Grey and Michael Rae, *Ending Aging: The Rejuvenation Breakthroughs that Could Reverse Human Aging in Our Lifetime* (New York: St. Martin's, 2007). De Grey calls his approach Strategies for Engineered Negligible Senescence (SENS). There has been tremendous controversy about de Grey's ideas, and MIT's *Technology Review* commissioned a competition for articles that might prove his theories so wrong as to be unworthy of learned debate. A panel of independent experts assembled by MIT chose three proposals, but none of them met the criteria of the challenge, thereby suggesting that de Grey's ideas are indeed very plausible on a scientific level.

48. Ray Kurzweil, *The Singularity Is Near: When Humans Transcend Biology* (New York: Penguin, 2006).

49. Jerold Schnoor, "Seven Ideas Lost on America," *Environmental Science and Technology* 42 (March 1, 2008): 1389.

50. K. C. Armitage, "The Continuity of Nature and Experience: John Dewey's Pragmatic Environmentalism," *Capitalism Nature Socialism* 14 (September 1, 2003): 49–72. See also Andrew Light, *Environmental Pragmatism* (Routledge, 1996).

51. For a scientific description of the term, see Stephen Jay Gould and Elizabeth S. Vrba, "Exaptation—A Missing Term in the Science of Form," *Paleobiology* 8 (1982): 4–15.

52. Kenneth S. Deffeyes, *Beyond Oil: The View from Hubbert's Peak* (New York: Hill and Wang, 2006), 169.

Epilogue: Embracing Uncertainty

1. Todd McCarthy, "Wall-E," *Variety,* June 26, 2008.
2. Patrick Ford, "Wall-E's Conservative Critics," *The American Conservative,* June 30, 2008.
3. Robert M. Hazen, *The Diamond Makers* (New York: Cambridge University Press, 1999), 147.

INDEX

Abandoned Mine Land fund, 198
Aboriginal Heritage Act (1972), 199
addictive behavior, 97–98
Adney, Tappan, 132
Affluenza, 93
Africa: AIDS as issue in, 145; ancient cities in, 64–65; conflict over diamond mining in, 45–46; conflict over oil in, 118–122; diamond mining in, 44–45; ritual significance of metals in, 25, 27–28; survey of priorities in, 227–228
Agarwal, Anil, 209
Agricola, Georgius, 136, 209–210
agriculture, 63–65
AIDS, 47, 145
Ajuwon, Bade, 27
Alaska: environmental concerns in, 133–134; gold rush in, 31, 68; oil extraction in, 128, 130; resource rights in, 132–133
Alaskan Native Claims Settlement Act, 132
Albertus Magnus, 3
Albright, Madeleine, 39
alchemy, 3, 5
Alderfer, Clayton, 106
algae, 219
Ali, Muhammad, 111
Allenby, Braden, 171–172
Ally, Russell, 76

aluminum, 23, 82–83, 168; recycling of, 166
amber, 53–55
American Chemical Manufacturers' Association, 174
ammonia, 100–101
Angola, 120
anorthosite, 207
Appalachia: coal mining in, 139–140; ginseng cultivation in, 198
Ararat, Mount, 22
Arendt, Hannah, 89
Argyle diamond mine (Australia), 60, 200
Arizona Enterprise Zone Program, 203
Arkansas: diamond mining in, 128
Arrow, Kenneth, 230
arsenic, 24
asbestos mining, 162
Asea Brown Boveri, 70–71
Ashanti Goldfields, 145
AT&T, 171–172
Athabascan oil sands, 84–85
atoms, 4
Australia, 21; aboriginal people of, 57, 199, 200; diamond mining in, 60, 200; land reclamation in, 200; pearling in, 57
Australia (film), 199
Austria: salt mining in, 34

automobile industry: in Japan, 175; and life-cycle analysis, 177
Auty, Richard, 115
Ayres, Robert, 172

Bainton, Roland, 29
Bairoch, Paul, 63
Bancroft, Hubert Howe, 68
Bangkok, Thailand, 29
Barak, William, 199
BASF, 100–101
Bauer, Max, 43
Baum, Frank, 36
bauxite, 151, 152
Bavaria: salt mining in, 34
Becker, Gary, 97
Becquerel, Henri, 5
Bell Laboratories, 171–172
Berenger, Henry, 119
Bernstein, Peter, 31
BHP Billiton, 156
Bilger, Burkhard, 122
Bingham Canyon mine (Utah), 137
Bisbee, Ariz., 202–203
bitumen, 80, 85
Blizzard, Reeze, 140
Blood Diamond, 46
Bodiva clay pit, 194–195
Bolivia: nitrates in, 100
Bonds, Judy, 140
Bosch, Carl, 100, 101
Botswana: AIDS epidemic in, 47; diamond mining in, 47–50; as stable economy, 112, 113, 127
Bowles, Samuel, 226
Boyle, Robert, 3
Brazil: diamond mining in, 43
Brickman, Philip, 107
brine, 32, 33
British East India Company, 43
Brodie, Richard, 91, 92, 94
bronze, 24
Bronze Age, 23, 24, 79
Bryan, Henry, 207
Budd, J. Danley, 207
Buddhism: diamonds as symbolic in, 51–52; gold as used in art of, 29

Burundi, 28
Bush, George W., 109, 146
Butchart Gardens, 194

Cable, Herman W., 207
calcite, 71
calcium, 4
calcium sulfate, 40, 41
California: gold rush in, 65–67
Cameroon, 118
Camminetti Act, 137
Campbell, Donald, 107
Canada: environmental reclamation efforts in, 187–188; oil resources in, 85
carbon: many forms of, 6–7, 41, 43–44, 78. *See also* coal; diamonds
carbon dioxide, 147
Carlsbad Reclamation Project, 185
Carnegie, Andrew, 210
Carter, Jimmy, 198
Çatal Höyük, 22
cellular phones, 162
cement, 40
Center for Disaster Control for Planet Earth (CDCPE), 236
cerussite, 1
Chad: conflict over oil in, 118–119, 121–122
chakras, 36
chaos theory, 94–95
Chevron-Texaco, 120
child labor: in coal mining, 144; in the diamond industry, 52
Childe, Vere Gordon, 64
Chile, 127; mining in, 145; nitrates in, 99–100
China: coal mining in, 79, 148–149; environmental issues in, 149–152; industrialization in, 147–148; iron mining in, 148, 151; pearls as venerated in, 56; population regulation in, 230; salt in, 32–33, 34; significance of minerals in, 36–38; steel production in, 148
chlorine, 31
Christianity: gold as symbolic in, 30–31
chromium, 6

Cisse, Youssouf, 25
cities. *See* urbanization
civilizations: collapse of, 19–20; and desire for adornment, 20–21. *See also* urbanization
Clark, Karl A., 85
Clement V, Pope, 29
climate change, 187–188
Clinton, Bill, 213
Clinton, Hillary, 128
coal, 7, 79
coal mining: environmental concerns relating to, 137, 139–140; hazards involved in, 142–144, 148–149; by Navajo and Hopi people, 217; reclamation efforts in areas damaged by, 197–198
Coal River Mountain Watch, 140
cobalt: in Congo, 111
Coeur d'Alene Basin, 187
coinage: metals used for, 2, 167. *See also* money
Cold War, 97
Collier, Paul, 112, 115, 116
coltan, 126, 162
Columbus, Christopher, 56
commerce: early forms of, 63–64. *See also* gold standard; money
Commoner, Barry, 171
Communities and Small-scale Mining (CASM), 152
computer recycling, 11
Congo, 27, 44, 45, 47, 109, 111, 112, 126
Connecticut: copper mining in, 197
Conrad, Joseph, 111
Conservation Reserve Program, 157
consilience, 5
consumption: as addiction, 97–98; in China, 152; decisions about, 95–96; environmental impact of, 101–102, 108–109, 230, 235–239; and happiness, 103–109; impact of, 93–96; and patriotism, 109; as political force, 96–99; and technological advancement, 182–183
Cook, James, 192
copal, 53

copper, 196, 209; in African societies, 27; in ancient societies, 23, 24; in China, 151; in Congo, 111; mining of, 6, 197; recycling of, 166–167
Copper Queen Mine, 202, 203
Cornwall, England, 194–195
corundum, 207
Costa Rica, 156
Crater Lake National Park, 122
Crichton, Michael, 55
crofters, 189
cryolite, 82–83
Crystal Palace (Leadville, Colo.), 12
crystals, 39–41
Csikszentmihalyi, Mihaly, 94
Cuesta, Efren, 40
cultural behaviors, 91–94
Curie, Eve, 5
Curie, Marie, 5–6
Curie, Pierre, 5
Curzon, Lord, 119
cyanidation process, 75
cyanide: and gold processing, 150
cybernetics, 170–171
Cyprus: copper in, 24
Cyprus Amax Corporation, 186
Czech Republic, 21

Dahle, Øystein, 157
Daily, Gretchen, 157
Daly, Herman, 225, 227
Darwin, Charles, 114
Darya-e-Noor Diamond, 43
Dasgupta, Partha, 153
David, Paul, 70
Davis, Graham, 117
Davy, Humphrey, 143
Dawkins, Richard, 91
De Beers, 49–50, 59, 60
Deffeyes, Kenneth, 233
De Graf, John, 93
de Grazia, Victoria, 96
de Grey, Aubrey, 231
Deming, W. Edwards, 176
Democratic Republic of Congo (DRC), 110–113
Denmark: landfill taxes in, 11

Denney, Reuel, 102
desalination, 32
Detroit, Mich.: salt mining in, 161–162
Dewey, John, 232
Diamond, Jared, 20, 84
diamonds, 7; chemical nature of, 43–44; colored, 60; in Congo, 111; cutting of, 51–52; mining of, 42–43, 44–45, 47–50, 128; social issues surrounding, 45–50, 52; as source of conflict, xii, 116; synthetic, 59, 60, 238
Diamond Sutra, 51–52
Diavik diamond mine (Canada), 60
DiCaprio, Leonardo, 46
Dickinson, Emily, 15
Dietrich, Wolf, 34
DNA: preserved in amber, 55
dopamine, 89–91
Drake, Edwin, 33
Dutch disease, 117
Dwan, John, 207
Dye, David, 146
dynamite, 71

Earle, Sylvia, 218
Earth Summit (1992), xi
Earthworks, 116, 128, 134
Easterly, William, 225, 227
ecocentric management, 180–182
eco-industrial parks, 174
ecological economics, 155–158
ecology, 170–171. *See also* environmental restoration; industrial ecology
economic diversification, 215
Ecuador: deforestation in, 81
Eden Project, 194–196
Edwards Woodruff v. North Bloomfield Gravel Mining Company, 136–137
Ehrlich, Paul, 217
Einstein, Albert, 6
electricity, 53–54, 217
elements: ancient Greek view of, 3, 36; Chinese view of, 36–38; conversion of, 5; periodic table of, 3–4; as used in manufacturing, 8
energy, 172–173
Engels, Friedrich, 142

entropy, 172–173
environmental concerns: in Alaska, 133–134; in China, 149–152; and coal mining, 137, 139–140; and coal-fueled electricity, 217; and consumerism, 101–102, 108–109, 230, 235–239; and the collapse of civilizations, 20; and economic growth, 225–226; and gold mining, 74, 134, 150; holistic approach to, 170–171; international efforts in response to, 140–141, 152, 156–157, 218; lawsuits involving, 136–137, 140–141; and oil extraction, 121; and phosphate mining, 154; and plastics, 8–9; and regulation, 229–230; and resource extraction, 134–141, 149–152, 186–188, 197–198, 231–234; and synthetic diamonds, 60; and uranium, 134–135; and zinc mining, 186. *See also* ecological economics; industrial ecology; sustainability
Environmental Protection Agency (EPA), 133, 186, 187
environmental restoration, 186–188, 204–206; in Australia, 200; and the Eden Project, 194–196; monitoring of, 205–206; in New Caledonia, 193; plants as used in, 195–196; in Scotland, 191–192, 196–197
Environment Canada, 195
Epicurus, 87
Equatorial Guinea, 119–120
euro, 73
evolutionary theory, 19
exaptation, 233
exergy, 172–173
Exodus, 28
Extractive Industries Transparency Initiative, 128
Exxon Valdez disaster, 133

Faragher, John, 66
Farah, Douglas, 46
Fearon, James, 116
feng shui, 37
Ferris, James, 18
fertilizer, 99

fiberglass, 162
Fiji, 203
filibustering, 67
fire: in smelting metals, 23–25
flint tools, 22
Forbes magazine's list of billionaires, 213
Ford, Gerald R., 198
Ford, Patrick, 236
Foreman, George, 111
fossil fuels, 78–82
Franco, Francisco, 40
Freedman, David, 114–115
Freeman, John, 179
Friedman, Benjamin, 225
Friedman, Thomas, 96, 118
Friedrich Wilhelm (Prussian king), 53
fuels. *See* coal; fossil fuels; natural gas; oil; whaling
Fuller, Buckminster, 195

Gabon: deforestation in, 81
Galbraith, John Kenneth, 155
Galton, Francis, 114
gamma-aminobutyric acid (GABA), 90
Gandhi, Mohandas, 34
Gangin, 136
garbage. *See* waste materials
Garcia-Guinea, Javier, 39–41
Garrett, Pat, 185
gas. *See* natural gas
Gates, Bill, 83, 212
Gbadolite, Congo, 111–112
Geber bin Hayyan, 3
Gemological Institute of America, 59
gemstones: allure of, 39, 41–42, 61; healing powers of, 36; sources of, 58–59; synthetic, 59–60. *See also names of specific stones*
General Electric, 59
geodes, 39–41
geomancy, 37
Georgescu-Roegen, Nicholas, 155
Ghana, 145
Giant Mine (Canada), 187, 201
Gibbs, J. William, 172
Gibney, Matthew, 57
Gilbert, William, 53–54

Gingrich, Newt, 229
ginseng, 198
Glazer, Nathan, 102
Gleick, James, 94
Glensanda estate, 190–191, 192
Glensanda Harbour Act, 190
Global Environmental Management Initiative, 174
globalization: and conflict surrounding resource management, 119–120, 225–226; and egalitarianism, 226
Global Witness, 116
Gods Must Be Crazy, The, 49
Golconda diamond mine (India), 43
gold, 2; in ancient Peru, 21; in Congo, 110–111; consumption of, 76–77; as currency, 30–31, 72; cyanidation process for, 75; hazards in mining of, 134, 144–145; in Indian matrimonial customs, 76–78; as investment, 73–74; mining of, 31, 65–67, 70, 71, 74–76, 150; new uses for, 219; recycling of, 167; religious significance of, 29–31; social issues surrounding, 74–76; trading of, 33; universal desirability of, 28–31, 72–74
Goldman Environment Prize, 140
gold rushes, 31, 65–67, 68
gold standard, 31, 72–73, 78
Golley, Frank, 170
Goppert, Johan, 44
Gould, Stephen Jay, 233
Graedel, Thomas, 171–172, 218
granite quarrying: in Scotland, 190–192
graphite, 7
Greeks, ancient: amber valued by, 53; elements as viewed by, 3, 36
greenhouse gases, 147
Grimshaw, Nicholas, 196
Guadalupe-Hidalgo, Treaty of, 66
Guerrero, Manuel, 40
gypsum, 40

Haber, Fritz, 100, 101
Haeckel, Ernst Heinrich, 170
Haggard, H. Rider, 35
Hannan, Michael, 179

happiness: and affluence, 102–109
Hardin, Garrett, 230
Harmony Gold, 145
health concerns: associated with mining, 142–145
heat exchange, 173
hedonic treadmill, 107
Herbert, Eugenia, 27–28
Hindu mythology: and diamonds, 42, 43; and minerals, 36, 37
Hine, Robert, 66
Hoeffler, Anke, 115, 116
Holm, John, 48
Hope Diamond, 43
Horgan, John, 231
Hotelling, Harold, 153
Hsu Chin-I, 38
Huber, Peter, 83
Hugo, Victor, 235
Humphreys, Macartan, 117
hypoxia, 145

Inco nickel-mining company, 193
India: diamond mining in, 42–43; diamond trade in, 42–43, 44–45, 51–52; environmental restoration in, 197; scrap metal in, 168; significance of gold in, 76–78
Indonesia: gas drilling in, 146
Indra, 42
industrial ecology, 171–173; and corporations, 179–183; critics of, 183; in Denmark, 173–174; implementation of, 174–179
industrial revolution, 169–170
International Chamber of Commerce, 174
International Convention for the Regulation of Whaling, 80
International Court of Justice, 210
International Institute for Sustainable Development, 220
International Labor Organization, 52
International Seabed Authority, 218
inverted quarantine, 98
IPAT equation, xi

iron, 8; in China, 148, 151; recycling of, 166; smelting of, 24–25, 27, 70; as source of nourishment, 18–19; in Sweden, 70–71; symbolism in smelting of, 27–28
Iron Age, 23
Islam, 17; gold as valued in art and architecture of, 29; pearls as venerated in, 56
itacolumite, 43

Jacobs, Jane, 63–64
jade, 37
Jahangir (Mughal emperor), 43
Jains: as diamond cutters, 51–52
Japan: auto industry in, 175; market for mushrooms in, 123; pearling in, 57–58
Japanese Ministry of International Trade and Industry, 171
Jevons, William Stanley, 182–183
Jevons paradox, 183
jewels. See gemstones; *names of specific stones*
Johannesburg, South Africa, 68
John Paul II, Pope, 33
Jones, Mary Harris (Mother Jones), 140
Jordan, Chris, 98
Journal of Industrial Ecology, 177
Jurassic Park (Crichton), 55

Kaaba (Mecca), 29
Kahneman, Daniel, 106
Kalahari desert: San people of, 48–50
Kaunda, Kenneth, 112
Kelvin, Lord, 113, 114
Kenya, 93; iron production rituals in, 27–28; political chaos in, 119
Keynes, John Maynard, 62
Khama, Seretse, 50, 112
Kieffer, Susan, 236
Kimberley Process, 45, 128
Kingairloch estate, 190, 191
King Solomon's Mines (Haggard), 35
Kinshasa, Congo, 110–111
Kirsch, Stuart, 141
Klee, Robert, 218
Klondike gold rush, 31, 68
Koh-i-Noor Diamond, 43
Koucky, F. L., 24

Kristof, Nicholas, 52
Kurzweil, Raymond, 231–232
Kuznets curve, 226

La Brea tar pits, 85
LaDuke, Winona, 134
LaFarge company, 193
Lakshmi, 37
Lave, Lester, 177–179
Lavoisier, Antoine, 3
lead, 4; mining of, 1–2, 6, 133
Leadville, Colo.: Crystal Palace in, 12; silver mining in, 1–2, 13–14
Legendre, Adriene-Marie, 114
Lemisch, Jesse, 103
Leonardo da Vinci, 33–34
Leopold, King, 112
Liberia: civil war in, 126
Liddicoat, Richard, 59
life-cycle analysis (LCA), 177–179
life cycles, 17–18; copper as symbolic of, 27; minerals as part of, 18–19
Life-Gem, 61
Linder, Staffan Burenstam, 95
Lion's Cave (Swaziland), 20–21
Loch Linnhe, 190, 191
lodestone, 35
Lomborg, Bjorn, 232
London, Jack, 31
Lonely Crowd, The (Riesman), 102
Lord of the Rings (Tolkien), 36
Lorenz, Edward, 94
Louis XIV, King, 43
Luderitz, Adolf, 135
Lumumba, Patrice, 112–113
Lupang Pangako, 163–164
Lyubomirsky, Sonja, 105

MacDonald, George, 35
McGonagle, William A., 207
McIntosh, Alastair, 191
McKibben, Bill, 226
McVeigh, Timothy, 99
Madagascar: sapphire mining in, 124–125; vanilla cultivation in, 123–124
magnesium, 4

Mahdavy, Hossein, 117–118
malaria, 145
Malaysia, 127
Mali: salt trade in, 33
Mallaby, Sebastian, 227
Malthusianism, xi
Mande people, 25
manganese, 218
Manila, Philippines: rubbish dump in, 163–164
Mao Zedong, 56, 148
Marcos, Imelda, 164
Marlowe, Christopher, 110
Marshall, James, 65
Martial, 55
Maslow, Abraham, 105–106
materialism. *See* consumption
materials: collecting of, 19; composite, 10; consumption of, 6, 8, 19–20; human quest for, 2–3; obsolescence of, 169; and weapons manufacturing, 6. *See also* consumption; minerals; plastics; waste materials
Matsutake mushrooms, 123
Mecca, Saudi Arabia, 29
Medina, Martin, 165
Mehta, Dilip, 51
Melbourne, Australia, 68
memetics, 91–94
Mendeleev, Dmitri, 4
mercury, 4, 150, 151
Meroe, 64
metals: in ancient societies, 23–28; as commodity, 167–168; recovery and recycling of, 166–168; rituals associated with, 27–28; smelting of, 24–25; as used in manufacturing, 7. *See also* copper; gold; iron; platinum
metalsmiths, 25
meteorites, 25
methane, 219
methyl hydrate, 86
Mexican-American War, 66
Meyer, Lothar, 4
Microsoft Corporation, 91
Mikimoto, Kokichi, 58

Miller, Daniel, 103–104
Mills, Mark, 83
Mineral Policy Center, 134
minerals: consumption of, 4; as currency, 72; deities associated with, 27, 37, 42; defined, ix–x; as factor contributing to conflict, 112, 115–118; as food displays in China, 38; in Hindu mythology, 36, 37; importance of in ancient societies, 22–23, 38; and life cycles, 18–19; in literature, 35–36; renewable sources of, 99–100; as source of nourishment, 4, 18–19, 36; underwater mining for, 218; from volcanoes, 22–23; wealth and philanthropy associated with, 209–214. *See also* gemstones; metals; resource extraction
Mine Safety and Health Administration (MSHA), 146, 147
mining: community veto of projects involving, 224–225; environmental issues surrounding, 134–141, 149–152, 186–188, 197–198; hazards of, 142–144, 148–149; hydraulic, 136–137; innovations in, 71, 75; and sustainability, 220–225. *See also* coal mining; copper; diamonds; gold
Mining, Minerals, and Sustainable Development (MMSD) initiative, 220–223
Miningwatch, 134
Minnesota Mining and Manufacturing (3M), 207–208
Mobutu Sese Seko, 111, 112
molybdenum, 166–167
Mombasa, Kenya, 93
money: early forms of, 71–72; in *Star Trek* stories, 214. *See also* gold standard
Mormonism: pearls as venerated in, 56
Morse, Kathryn, 31
Morvern peninsula (Scotland), 190
Moseley, Henry, 4
Moses, 28
mosquitoes, 145
mountaintop-removal mining, 139–140
Mumford, Lewis, 109

Murphy, Kevin, 97
mushroom hunting, 122–123

nacre, 56
Nader, Ralph, 103
Namibia, 135
NANA Regional Corporation, 133
Naples, Italy: waste management in, 165–166
Nash, June, 141
Nast, Thomas, 30
National Industrial Symbiosis Programme, 174
Native Americans: salt as valued by, 31
natural gas, 79, 80
Nauru, 153–154
Navajo territory: coal mining in, 217; uranium mining in, 135
Ndinga-Mbo, Abraham, 25
needs: hierarchy of, 105–106; versus "wants," ix
Neil, Cecily, 203
New Caledonia, 192–193
Newton, Isaac, 43–44
nickel mining, 192, 193
Nietzsche, Friedrich, 17
Niger, 27
Niger delta, 120, 121
Nigeria, 120–121
niobium, 9
Nishikawa, Tokichi, 58
nitrates, 99–100; synthetic, 100–101
nitrogen, 100, 101
Nobel, Alfred, 71
Nobel Prizes, 5–6, 101
nonrenewable resources, 9–10
Nordhoff, Charles, 70
Norway, 130
Norwood, Charles, 146
nuclear reactions. *See* radioactivity
nylon, 99

Oastler, Richard, 144
Obiang, Teodoro, 120
obsidian, 22–23, 63
obsolescence, 169

ocean: plastics in, 8; as source of salt, 32
ochre, 21
Ogun, 27
oil (petroleum), xii; early discovery of, 33; as factor contributing to conflict, 116–122; as fuel, 7, 79–81; prices of, 81; supply and demand for, 81–82; from whales, 80
oil sands, 84–85, 86
Oldenburg, Veena, 77
opals, 59
Oppenheimer, Nicholas, 49
Ordway, Lucius, 207
organic chemistry, 7, 78. *See also* carbon
organic compounds, 7–8
Organization for Economic Cooperation and Development, 157
Oxfam, 116, 128

Papua New Guinea, 140–141, 218
Pareto, Vilfredo, 230
Pareto optimality, 230
Pauling, Linus, 6
pearls, 55–58; cultured, 58; diving for, 56–58; formation of, 55–56; freshwater, 56
Pecos River, 185–186
Pennsylvania Coal Company, 137
Pérez Alfonzo, Juan Pablo, 147
Peru: gold in, 21
pesticides, 149
Peter the Great, 53
petroleum. *See* oil
Pew, James Howard, 85
Pew Charitable Trusts, 84
pewter, 2
Philip IV (king of France), 29
philosopher's stone (*lapis philosophorum*), 3
phosphate mining, 153–154
phthisis, 143–144
phyllotaxis, 196
plants: cleansing role of, 195–196
plastics: manufacture of, 6; in the ocean, 8
platinum: recycling of, 167; in South Africa, 127
Plato, 3
Poland: salt mining in, 33–34
Polk, James K., 66

pollution. *See* environmental concerns
Polo, Marco, 32–33
polylactic acid, 9
Pope, Alexander, 54
population: increases in, 6; regulation of in China, 230
Porter, Theodore, 114
Post-It notes, 208
potassium, 4
Princen, Thomas, 108
Princess and the Goblin, The (MacDonald), 35
Projet de Gouvernance des Ressources Minérales (Project of Governance of Mineral Resources, or PGRM), 124–125
protons, 4
Putnam, Robert, 102
pyrite (fool's gold), 70

Al Qaeda, 46
qi, 37
quantitative analysis, 113–114. *See also* regression analysis
quartz, 43

radioactivity, 5. *See also* uranium
Raffaelli, Jean-François, 165
Rant, Zoran, 172
Rathje, William, 168–169
Ratho Center, 196–197
recycling: of metals, 166–168; of computers, 11. *See also* industrial ecology; resource recovery
refining process, 139
regression analysis, 114–115, 118
regulation, corporate, 229–230
remuneration, differentials of, 226–227
rentier states, 117–118
resource-curse hypotheses, 112, 115–118, 126–131
resource extraction: chronicles of, 135–136; environmental concerns surrounding, 134–141, 149–152, 186–188, 197–198, 231–234; as factor contributing to conflict, 112, 115–122; impact of, 183–184; and "optimal rate of depletion," 153; and propensity for

underdevelopment, 112–113; scenario analysis of impact of, 220–225; and sustainability, 215–219. *See also* industrial ecology; sustainability
resource recovery, 162–163, 165–169, 177–179. *See also* industrial ecology
resource-war determinism, 215–216
Responsible Care, 174
Revenue Watch Institute, 128
Rhodes, Cecil, 213
ribonucleic acid (RNA), 18
Richards, Leonard, 67
rickets, 144
Riesman, David, 98, 99, 102
risk-taking behavior, 92, 213–214
Roach, Geshe Michael, 52
Rockefeller, John D., 36, 210, 212
Roddenberry, Gene, 214
Rogoff, Kenneth, 227
Romero, Angel, 40
Roosevelt, Theodore, 185
Rose, Robert E., 207
Ross, Michael, 118
Royal Bafokeng Nation (South Africa), 127
Royal Oak Corporation, 201
Russell, Bertrand, 103, 104
Russia, amber mining in, 54
Rutherford, Ernest, 5

Sachs, Jeffrey, 115
Sacramento, Calif., 68
safety lamps, 143
Sagoff, Mark, 156
Sago Mine tragedy, 146
salt: mining of, 33–34, 161–162; as preservative, 32; as political issue, 34–35; as valued commodity, 31, 32–34
Saltpeter War, 100
Salzburg, Austria, 34
San Jose, Calif., 65–66, 67
Sanderson, Eliza, 137
Sanderson, J. Gardiner, 137
sandstone, 43
San people (Botswana), 48–50
sapphires: in Madagascar, 124–125

Saudi Arabia, 17, 29; state ownership of oil resources in, 130
Sawyer, Lorenzo, 136–137
scavenging, 164–166
Schaeffer, William, 95
Schleiden, Matthias Jakob, 115
Schnoor, Jerald, 232
Schor, Juliet, 108–109
Schumacher, E. F., 155
Schumpeter, Joseph, 155
Scottish Highlands, 188–190; land restoration in, 191–192; mining and quarrying in, 189–192
scrap metal, 166; from the World Trade Center, 168
Seabrook, John, 166
seashells, 22–23
selenite, 40
self-actualization, 105–106
Sen, Amartya, 230
Seneca, 87
Seurat, Georges, 98
Shakespeare, William, 159
Shapiro, Judith, 148
Shelley, Percy Bysshe, 102
Shen Nong, 36
Sherman Silver Purchasing Act, 2
Shirley, Joe, 135
Shrivastava, Paul, 180
Shuster, Joe, 41
Siberia: diamond mining in, 42
Siddhartha, 29
Siegel, Jerry, 41
Sierra Leone, 113
Silicon Valley, 65, 67
silver: mining of, 1–2, 10, 133; retrievability of, 10–11
Simon, Julian, 217
Simonyi, Charles, 91
Singh, Dhiru, 52
slag, 25
slavery, 125–126; as issue in California, 67
Smith, Adam, 104, 115
Smith, Cyril Stanley, 21
Smuts, Jan, 170
sodium, 4, 31. *See also* salt
Solomon, King, 35

Solow, Robert, 175
South Africa: diamond mining in, 42, 76; gold mining in, 134; platinum mining in, 127
Soviet Union: consumer demand in, 98–99
"Spaceship Earth," 195, 230
Spain: geodes in, 39–41
Specter, Arlen, 146
Spooner, Henry, 165
Sri Lanka, 89, 90
Standard Oil, 36
steady-state economics, 225
steel: in China, 148; recycling of, 166
Steinbeck, John, 56
Sterne, Laurence, 161
Stiglitz, Joseph, 227
Strong, Herb, 238
strong sustainability argument, 9
sugar, 125
Sumerians, 36
Sun Oil Corporation, 84
Superfund cleanup sites, 186, 187
Surat, India, 125; diamond trade in, 42–43, 44–45
Surface Mining Control and Reclamation Act, 198
Suriname, 156
sustainability, 214–219; and the mining industry, 220–225; physical, 214–215; scenario analysis of, 220–225; and social concerns, 215. See also industrial ecology
Sutter, James, 65, 68
Sutter's Mill, 65, 66, 67
Swahili language, 93
Swaziland, 20
Sweden: iron industry in, 70–71
Syncrude Canada, 85
Szasz, Andrew, 98

tailings, 139
Tansley, Alfred, 170
tar, 85
tar sands, 139
Tavernier, Jean-Baptiste, 43
Taylor, David, 196

tea, 89–90
technological advancement, xi–xii; and consumption, 182–183; and exaptation, 233; limits to, 232
Teck Cominco, 134
Tennant, Smithson, 44
Terra Amata (France), 21
Terrero Mine, 186–187, 188
terrorism: diamonds and funding for, 46
Thales, 53
Thomas process, 70
Three Gorges Dam, 85
3M Corporation, 207–208
Thutmose IV, Pharaoh, 28
Timaeus (Plato), 3
Timbuktu, Mali, 33
tin, 2, 24
Tolkien, J. R. R., 35–36
total quality environmental management (TQEM), 176–177
treasure impulse: in ancient societies, 21, 22–23
Triffin, Robert, 78
Trollope, Anthony, 136
Trout Unlimited, 186
Twitchell, James, 103

Union Carbide plant (Bhopal, India), 180
United Nations Development Programme, 149
United Nations Environment Programme, 115
U.N. human development index, 47, 48
United Nations Industrial Development Organization (UNIDO), 75
United States: diamond mining in, 128; mushroom hunting in, 122–123; oil wealth in, 128, 130
U.S. Department of Agriculture, 157
Updike, John, 1
uranium: in Congo, 111; mining of, 134–135; recycling of, 166–167
urbanization, 62–65; mineral resources as factor in, 63–64, 65–71
Usmanov, Alisher, 213
Uys, Jamie, 49

van der Lingen, Elma, 219
vanilla cultivation, 123–124
Varela, Janice, 186
Vatakoula Gold Mine, 203
Vedanta Resources, 209
Vedanta University, 209
Vedic scriptures, 36
Venezuela, 130, 147
Verneuil, Auguste, 59
Verney Report, 190
Virgin Mobile, 162
Vogel, David, 229
volatile organic compounds (VOCs), 208
volcanoes, 22–23

Wahhabi sect (Islam), 17
Wall-E, 235–236
Wal-Mart, 128, 179, 210
Walton, Sam, 128, 210
Ward, Ken, 147
Warde, Ibrahim, 46
waste materials, 162–163; from electronics, 168; scavenging of, 164–166; value of, 168–169. *See also* industrial ecology
water: quality of, 204–205; reclamation of, 185–186
Wat Traimit temple (Bangkok), 29
weak-state hypothesis, 116–117
weak sustainability argument, 10

wealth: early forms of, 71–72; gold as measure of, 72; and happiness, 102–109; and philanthropy, 209–214
weapons: manufacture of, 6
Weisman, Alan, 23
Wentorf, Robert, Jr., 238
whaling, 80
Wiener, Norbert, 171
Wilde, Oscar, 103
Wilson, E. O., 5, 192
Wilson, Woodrow, 96
women's rights: in the Middle East, 118
World Bank, 227
World Gold Council, 73
Worldwatch Institute, 116
World Wildlife Fund (WWF), 157, 193
Wright, Gavin, 70
Wunder, Sven, 81

Yantarny amber mine, 54
Yap, 71–72
Yeoman, John, 190
Yoruba culture, 27
Yunus, Muhammad, 214

Zambia, 112, 113
Zigong, China, 33
zinc, 8, 133, 166–167, 186
Zoroastrians, 17–18